Hot, Hungry Planet

HOT, hungry PLANET

The Fight to Stop a Global Food Crisis in the Face of Climate Change

Lisa Palmer

St. Martin's Press New York

Printed in the United States of America. For information, address
St. Martin's Press, 175 Fifth Avenue, New York, N.Y. 10010.

All photographs are courtesy of the author except where indicated.

www.stmartins.com

Design by Meryl Sussman Levavi

Cataloging-in-Publication Data is available from the Library of Congress.

ISBN 9781250084200 (hardcover)
ISBN 9781250096395 (e-book)

Our books may be purchased in bulk for promotional, educational, or business use. Please contact your local bookseller or the Macmillan Corporate and Premium Sales Department at 1-800-221-7945, extension 5442, or by e-mail at MacmillanSpecialMarkets@macmillan.com.

First Edition: May 2017

10 9 8 7 6 5 4 3 2 1

Dedicated to my family

Contents

Acknowledgments

I am one fortunate author. Not only is this my first book on a sub-ject about which I care deeply, but it's a volume that holds hope in stopping a global food crisis in the face of a changing climate.

I owe my greatest debt to Ed Purpura and to my sons and daughter and dogs for keeping me sane and relatively stable. My son Aidan demonstrated the value of a liberal arts education with his innate talent for editing, and he has a bright career ahead of him. My dearest friend, Donna, swooped me up early each morning for long runs to help me avoid getting antsy at my desk.

I am grateful to my warm and loving neighbors, friends, and colleagues at the National Socio-Environmental Synthesis Center (supported by an award from the US National Science Foundation, grant #DBI-1052875) for looking after my heart and my head as I was finalizing the book. I thank the Woodrow Wilson Center and its Environmental Change and Security Program for a yearlong fellowship that guided the development of this work, and my assistant, Breanne. As well, I thank the Vermont Law School's Environmental Law Center, the Society of Environmental Journalists, the Solutions Journalism Network, and the Pulitzer Center for Crisis Reporting for providing critical travel support, professional development, and intellectual guidance.

Some of the reporting for this book appeared in similar or altered form in *Yale Environment 360*, *Climate Connections*, *Slate*, *The Guardian*, and *Nautilus*. I'm grateful to all these publications; a special thanks for their meticulous editing to Fen Montaigne and Roger Cohn, and to Bud Ward for patience and guidance.

I can't thank enough my literary agent, Jessica Papin, my skillful editor, Emily Carleton, and her superbly able assistant, Annabella Hochschild, for their vision and support in developing this book.

My travels for this book took me to iconic American farms and to places where few Americans venture, and I was always treated with kindness and generosity. I'm especially grateful to researchers and scientists for their unfailing courtesy and forbearance, and to the bighearted farmers of the Valle del Cauca in Colombia for their excellent meals. In India, the curry and masala chai were marvelous. In California, organic fresh vegetables, fruit and fish had new meaning for me.

My deepest gratitude goes to the people the world over, farmers, producers, development and aid workers, researchers, writers, and others who continue to work in this immensely rich cosmos.

Introduction

The Fight to Close the Food Gap

In June 2013, I suddenly found myself having a meal in the middle of a field with a scholar from Spain, an environmental lawyer who specialized in water resources, an expert on energy, and an ecologist. I had been awarded a fellowship to attend a Vermont Law School summer session course on the global challenge of feeding the world while protecting forests and other natural landscapes, and that evening I had gone with other fellows, visiting faculty, and staff to the home of John Echeverria, a professor of law who was then the acting director of the Environmental Law Center. The group had come to the summer law program for a range of reasons—to teach, to learn, to interact with others outside our professions—and the Echeverrias' gathering, an informal buffet dinner on their porch, was meant to encourage camaraderie among the faculty and students. Halfway through the meal I stood up when I heard what sounded like a lawn tractor approaching, its driver hollering. I looked to the front yard and saw a picnic table on wheels. "Are you coming?" the enthusiastic driver asked. At John's urging a few dinner guests joined me in nodding yes. The driver of the all-terrain table introduced himself as John's neighbor and friend. He was on his evening joy-ride and had hoped to find a few companions. Both John and the driver suggested those of us new to Vermont take our plates with us for the ride down to an alfalfa field to behold the green landscape.

"Which one?" I asked with curiosity.

"The one next to the river," he said and pointed to the river valley that stretched beyond the front yard. For a moment I hesitated. I was enjoying the historic Victorian farmhouse. The driver prodded: "When have you ever been on a motorized picnic table?" Never. So I joined in.

It was an amazing ride. We bounced along a tractor path through the field until we rested on a bridge over the Ompompanoosuc River.

The meal was entirely local, prepared with seasonings that gave the food a Latin twist by John's wife: locally produced pork, squash and beans from a nearby garden, and lettuce and herbs from the Echeverrias' backyard. Exhortations to "eat local" and "vote with your fork" have become ubiquitous since the mid-2000s, and the bounty before us was enough to make anyone a proponent. In this part of

Vermont local food is plentiful, and state policies support and encourage small-scale farming. But could states everywhere—and countries everywhere—support policies that encourage their local farms to grow abundant amounts of food for local populations? During this particular idyll in the alfalfa, it was tempting to hope that Vermont—its policies and local food culture writ large—might have the answer.

We are on the cusp of a global food crisis. But you may not know it if you are looking at Vermont. Parts of Vermont could best be described as the "eat local"–utopia. In much of the United States—and the world—city sprawl, among other issues, means that meeting citizens' calorie needs with local production is virtually impossible. And, in just two decades, an additional 2.6 to 4 billion people will be sitting down at the global table wondering what's for dinner, what is dinner, or even if we have dinner.[1] That's the equivalent of adding the population of New York City to the world's grocery lines every month for the next thirty-five years.

As we sat in the field, our moving dinner party discussed how a growing population is putting pressure on the world's water, land, and natural resources like never before. Planning ahead to address this fight to feed humankind is both a numbers game and an urgent social crisis. Calories, climate change, and acreage for farming are some factors on one side of the equation. The 7 billion–plus people on the planet now, projected to swell to 9.6 to 11 billion by 2050, are on the other.

How is the global food system meeting the demands of people right now? Of the more than 7 billion people in the world, about 1 in 6 go to bed hungry every night. This is not because we don't have food. This is not because we do not grow enough food. It is, for the most part, because about a billion people, or somewhat fewer, don't have the financial, institutional, or political means to get it. They're too poor. They're too disenfranchised. They're too disconnected from world affairs to exercise power to get this food. They are food insecure. Essentially it's a problem of poverty and institutions and not one of agronomy or land-use change or forests.

But in the near and far future, a growing global population with changing tastes will add to food insecurity, putting additional pressure on the food system. More important than population increases is dietary change in the rising middle class. Yes: adding more than 2 billion people to the planet during the next two to three decades is a big issue—it means a whopping 30 percent increase in food demand. But consider the increasing wealth of the world, especially the 4 billion people who are now becoming part of the global middle class. That group will increase from 1.8 billion in 2009 to 3.2 billion by 2020 and 4.9 billion by 2030. Most of this growth will come from Asia. By 2030 people in China, Indonesia, Vietnam, and other Asian countries will constitute 66 percent of the global middle-class population and account for 59 percent of middle-class consumption, compared to 28 percent and 23 percent, respectively, in 2009.[2] They are changing their diets and adopting a more Western style of diet. So that means eating more meats, more dairy products, more sugars, fats, oils, and other resource-intensive foods. Population growth is part of the issue, but more important is consumption: how diets will change with increasing affluence. There may now be sufficient food, if it is not wasted, but it is not always affordable. This illustrates a basic paradox of the food supply—once people have sufficient funds to afford food, they almost immediately want better food, which puts greater strain on the food system.

Vermont has had a long history of community-supported agriculture, an alternative to supermarket shopping where customers pledge to support a farm or group of farms and share the risks of farming with the growers. While the concept of community-supported agriculture originated in Japan in the 1960s and was further developed in the 1970s in Europe, farmers in New England—and especially Vermont's dairy cooperatives—have been moving toward closer connections to consumers ever since organizing themselves through cooperatives for the better part of a century.[3] Community-supported agriculture differs from a co-op, which introduces a middleman. In community-supported agriculture, consumers buy directly from their local grower without going through a middleman. The number of people buying their food directly from farms has grown steadily in response to the increasing demand for natural and organic food, particularly from the burgeoning urban population.[4]

Today, there are between 2,600 and 6,000 farms that sell to consumers through these direct agreements across the

United States. (People disagree on official ag census statistics.)[5] Yet some research provocatively suggests that local farming, as well as organic farming, is not always better for the environment than large-scale production for larger markets because small and organic producers often have lower yields and need more land to produce the same amount of food, resulting in more deforestation and biodiversity loss, which ultimately can undercut the environmental benefits.[6] Depending on the crop, consuming local food does not always translate to more efficient use of water, energy, and land resources than consuming food produced on large, modern farms for mass distribution. This is to say, complexities exist, and "buy local" isn't a silver bullet. When both small- and large-scale systems take on more elements of agroecology—a field that takes a whole system approach to agriculture production and combines biodiversity and ecology with farmers' knowledge and consideration of social and economic conditions—they can reduce agriculture's impact on climate change and make it possible for ecosystems to produce abundant sustainable food while also improving social and economic resiliency in food systems. Because feeding the world sustainably is a critical human problem rooted in the multifaceted and deeply interconnected relationships between humans and the environment, the response requires us to examine the entire food system holistically, from production through processing, distribution, and consumption. Ultimately the solution to feeding the hot, hungry teeming planet of the future is not in the rolling hills of the Vermont countryside, where ecological abundance and economic wealth create the possibility for delicious local dinners next to clean, flowing rivers. Rather, there will be many solutions in many different places that must reflect the relationships between global warming and global hunger, between the natural environment and the social environment, in our attempts to address both problems.

Local food has benefits and tradeoffs. It's almost always fresher than food that has been trucked 1,000 miles, but distance doesn't necessarily correlate with the environmental costs of that food. Writing in *World Watch Magazine*, Sarah DeWeerdt argues:

> If the goal is to improve the environmental sustainability of the food system as a whole, then there are a variety of public policy levers that we can pull. To be sure, promoting more localized food production and distribution networks would reduce transport emissions. But what if a greater investment in rail infrastructure helped to reverse the trend toward transporting more food by inefficient semi-truck? What if fuel economy standards were increased for the truck fleet that moves our food? Or, to name one encompassing possibility, what if a carbon-pricing system incorporated some of the environmental costs of agriculture that are currently externalized? Local food is delicious, but the problem—and perhaps the solution—is global.[7]

Researchers have found that 83 percent of emissions occur before food even leaves the farm.[8]

I worked hard to keep all this in mind while enjoying the delicious dinner in the alfalfa field in Vermont. The other guests and I swapped impressions about the impact of farming on ecosystems, the effects of extreme weather and drought on crops, and the destruction of forests and degradation of pastures to support a growing population that is eating more meat. We talked about fracking and nuclear power, water and soil, science and technology, and what kind of knowledge people might need in the future. As I looked around the table, my strongest impression was that the answer depended on all these things at once: social sciences like economics and politics, the environmental sciences, and the unequal capacities for adaptation; human geography and development; the rural and urban poor; health care and population; education and women's empowerment; and land use and technology.

When it comes to food, transportation contributes a tiny fraction of the overall environmental impact. Local food has plenty of benefits, and I'm all for it, but if you want to "think global," you need to consider other factors: changing our consumption of animal products; reducing human fertility rates, which is the average number of children that would be born per woman; expanding agriculture on treeless, degraded lands rather than within forests; and addressing the gross overuse—and underuse—of fertilizer.

In other words, Vermont's kumbaya agrarian model is not the global answer to how to feed the world. It might be the answer to feeding Vermont. Or part of Vermont. What we need is a sustainable and resilient global agricultural system. How we decide to feed the next 2.5 billion people will define civilization for millennia to come. Technically, farmers today grow enough food to feed everyone. Yet nearly 1 billion people periodically go hungry, usually because the food is too expensive or not available in the right places. Families in poorer parts of the world have historically consumed one meal a day. In some areas families plan their meals for the week and decide which two days they will go without eating entirely. The number of people who use this strategy will likely increase as the world population grows amid a changing climate.

So, maddeningly, our current food problem isn't linear. A major trend is more meat and dairy consumption, and it is affecting food production, land use, water use, and food prices globally. The rapid rise of the global middle class is driving half the increase in the world's predicted food consumption. To prevent more hunger farmers would have to more than double their production by 2050, even though the population will not have doubled. As more people gain wealth, they adopt diets that include meat and dairy. And producing a pound of chicken or beef takes four to twelve pounds of grain, respectively. So we run into a shortage—or destabilization of prices—as the world tries to produce more grain for animal feed.

As I sat in the Vermont field observing the richness of the land around us, the discussion turned to the expansion of farming. Someone wondered why we can't grow more food by planting more acres. Straightforward as this seems, increasing the amount of land devoted to agriculture is not the answer. Half the world's vegetated land is already devoted to agriculture. Clearing more forests, especially tropical forests, and grasslands would cause an environmental disaster. Nor are crop yields improving fast enough. To keep up with projected food demands, farmers will need to produce 2.4 percent more each year—and every year. Even with the spread of modern farming methods, farmers achieve half that, or a production increase of 0.9 to 1.6 percent per year. Finally, climate change is predicted to reduce crop growth by at least 1.5 to 2.5 percent per decade.[9] Climate change is forecast to significantly hurt some crops in some areas. In India, for example, research shows that wheat yields will suffer losses of 6 to 23 percent by 2050.[10]

As a journalist my job is to talk with leading experts to understand how society is working to address the food gap. While I was interviewing people on food production and climate change for this book, everyone I talked to seemed a little panicked, to say the least. I asked a plant researcher about how to improve rice yields enough to meet the food needs of a growing population under climate change. I was expecting a technical description of innovations in rice breeding or growing methods. His answer? Educate women and girls while giving them access to family planning in order to reduce population growth and the demand for food.

When I spoke with one of the world's leading experts

on global land use, I learned his solution to feeding a hot, hungry planet isn't improving soils or yields or preventing deforestation; rather, he pointed to the need to improve the health and nutrition of infants and children younger than five in many developing countries because their mental and physical capacities are stunted permanently as a result of being starved for calories and vitamins essential to growth.[11]

When I talked to a lawyer who develops policies on global agriculture, I learned her solution to feeding the world is to grant women legal rights to the land they are farming. And when I talked to a top authority on food policy in the developing world, he did not discuss his area of expertise—the use of genetically modified seeds, for instance—but rather the need to build better roads and better access to information technology through mobile phones. During all these interviews I was fascinated to note that the answers from experts did not always reflect their areas of expertise.

This was my eureka moment: As their responses made clear, addressing the fight to feed the world population will require disparate systems working together. The components of the global food system are interconnected, and these connections—especially those components and activities in one part of the system that strive for positive impacts across the system—are at risk. Furthermore, change within and around food systems is not likely to occur soon enough or on a scale large enough to avert disaster. We have begun to exhaust our planet and the natural resources that form the basis of the food system. We also have made social, economic, and policy decisions that do not always take into account unintended consequences and tradeoffs. However, as the experts also made clear, integrating ideas from both the social and natural sciences will act as levers and will effect positive change. When these levers are pulled simultaneously, the approaches that integrate education and empowerment, data and insights, policy shifts, disruptive technologies, and wise choices and politics can target root causes, address fundamental holistic solutions, and bring sweeping change to close the food gap. This book shows us how.

⊕

When I was growing up in Minnesota in the 1970s and '80s, my parents pretty much knew the sources of the main foods we ate. Eggs came from the egg man down the street; meat was from a side of beef they bought periodically from the meat locker in the tiny village of Bergen. Pork came from the seasonal hog roast and was frozen, or often from a hog farm a couple miles away. The local dairy supplied milk. Our bread and rolls came from the bakery on Main Street. Our family garden plot at my grandparents' farm, which had soil so black it looked blue, produced our vegetables. In winter our mom served canned or frozen vegetables from the summer and fall harvests. We did not eat much fruit. Sometimes my mom's cousins or friends brought us apples from their orchards. In proper 1970s fashion we had our share of processed foods like Wonder Bread, Campbell's tomato soup, Spaghetti-O's, Kraft macaroni and cheese, hot dogs, and bologna. But, at least while I was a young child, most of our sustenance came from local and farm-grown foods. That changed during my teens. Our family's dependence on packaged foods grew, and thanks to the industrial food system, it was inexpensive and easily accessible.

I witnessed the change in the U.S. food system in my own backyard. My first summer job was hoeing the weeds out of the soybean fields that surrounded my town of Jackson, Minnesota. Back then teenage labor—not the toxic herbicide Roundup—was the preferred method of weed control. In 1976 those fields in Jackson County produced twenty-eight bushels of soybeans per acre. By 2015 farmers were coaxing 47.5 bushels from the same area.[12] Yields increased over time, but now the growth in crop yields is stagnating. Chief among the concerns is the sustainability of staple crops such as soybeans, corn (maize), rice, and wheat. The world has been consuming corn and soybeans at a rapidly growing rate because more people are using them for biofuel and eating animals that are fed corn and soybeans. This is the direct result of the growth of the middle class worldwide.

Any successful answer to feeding a hot, hungry planet will result from continual change, shifts, and adaptation to an uncertain future. Figuring out where society can sustainably grow more food and how people might try to do that under climate change will be inextricably linked to national security, public health, and economics. In many ways the world's farmers and food producers are no more masters of their fate in their interactions with the environment and global markets than are hikers trying to summit a mountain. We may not have absolute control of our future, but we also have not left the future to fate. A skilled guide, supportive backpack and boots, and well-secured provisions can help guarantee a successful journey. In *Hot, Hungry Planet*, you will meet the skilled guides to a sustainable future. As their experiences and stories make clear, leveraging the changes in science and technology must be accompanied by improvements in education and women's rights, data and insights, policy shifts, and wise choices to meet the food needs of our planet.

Why is it so important that all these disparate systems—science and economics and education and agronomy, for instance—work together to ensure we have enough crops and livestock to feed the world? For most of our existence human populations have been small and the earth relatively large. One billion people were living on Earth as the nineteenth century became the twentieth. We had vast resources relative to the size of our society. That is no longer the case. We are now much more numerous as a species—more than 7 billion of us, hurtling toward 9 to 10 billion by the middle of the century. Population growth has been a big factor in this pressure on our planet. But perhaps even more important is the way we use resources and technology.

Our population has more than doubled since the mid-1960s. The world economy grew about sevenfold during the same time, and as a result global food and water consumption has tripled, and fossil fuel combustion has quadrupled. We've reached an inflection point in the history of this planet, and we'll need to effect change—quickly—if we want to stay here.

In general humans are better off than in the past. We live longer lives. We live healthier lives with better nutrition and better medical care. More of us live in democratic societies with greater literacy and freedom. We have more capacity to adapt our lives to changes around us. On one side of the equation humans are making tremendous progress. But on the environmental side things are not so good. We are exhausting water supplies. We are exhausting soil nutrients. We are exhausting our forests and biodiversity.

Three broad trends—two are troubling, and one might be harnessed for a better future—are shaping what comes next. The first, as I discussed earlier, is a growing global middle class hungering for more meat and dairy. The second is environmental degradation from the clearing of grasslands and forests to graze cattle, grow oil palm trees, and raise more grain for animal feed. Developing countries continue to clear vast forested areas at a remarkable clip. This has had devastating effects on human health, economics, land, ecosystems, biodiversity, and climate change.

The third trend is the one that gives me hope. More and more people are beginning to understand the need for economic opportunities to alleviate poverty. For 75 percent of rural poor, agriculture is the primary means of supporting their existence.[13] In the hierarchy of the rural poor, women occupy the lowest position. They comprise roughly 43 percent of the agricultural workforce worldwide, and they are the majority of agriculture workers in South Asia and sub-Saharan Africa.[14] Yet in many places women do not have rights to the land they farm, much less to a high school education. They don't have career options other than marriage, and they don't have the right to plan and space the births of their children.

In this book I will examine all three trends but especially the last one: the increasing awareness that social and environmental solutions are interconnected and can have positive feedbacks and impacts on multiple parts of the food system. Education and empowerment for women, the rural poor, and other marginalized groups that contribute much to local and regional food systems can open the door to profound social change while reducing both hunger and the environmental effects of agriculture. By placing particular emphasis on the rights of women and balancing the growth in the demand for food with the economic and social aspects of food security, we can address and reverse challenges posed by growing human populations and rising global temperatures in ways that positively reinforce other, related solutions.

Every week I hear about some innovation for helping women, but one program, developed by a woman in Uganda, truly exemplifies the complex and wonderful cascade effect. Let me tell you about Safira.

HOT

Women's Work

One humid sunny afternoon, an eighteen-year-old woman named Safira Nyesige was tending her family's new banana and maize plots with her mother when a neighbor arrived and implored Safira to discuss the latest farming innovation. What followed was a familiar series of events. Safira and a few other young women gathered to give a demonstration of new agriculture techniques they had learned at boarding school in Uganda. Some neighbor women asked about how to make mud bricks, others asked about seeds and soil. Before long dozens of villagers had gathered and were listening intently. Over the months the demonstrations grew so popular that Safira and her classmates began appearing on local FM radio broadcasts.

At school the students learned how to extend their classroom studies by working in partnership with their families. Safira did much of her homework during semester breaks and the summer holiday by farming the land with her mother and siblings. During her first visit home, when she was thirteen, the family built a latrine. A year later they planted corn, banana trees, and sweet potatoes and dug a pit for the family's garbage. Safira made bricks and began to raise chickens. By the time she was fifteen, she had moved her mother, three sisters, and four brothers out of their tiny thatched hut and into the brick house they had built next to it. The new home even had a corrugated metal roof, a luxury outside Kagadi Town, paid for with earnings from raising crops. The old hut became their kitchen. Outside the kitchen Safira built a new drying rack so pots, pans, and dishes could dry hygienically in the sun.

Home life has changed significantly for Safira. Before she attended school, her family ate one meal a day and sometimes not at all. They were frequently ill, she said. Home sanitation practices were crude. None of the children attended school. The family subsisted on the produce grown on their farm, but there was nothing extra to sell.

"We had poor health, poor education," she said but did not elaborate. Typically, families in the Kibaale District of western Uganda, where Safira lives, sent only boys to school. And as in many other parts of the world, girls not enrolled in school married early. But increasingly in Kibaale, early marriage comes with a stigma. "Girls who do not go to school have nothing to do with jobs," Safira said. "They end up being idles in villages and beggars on streets." Earning an income, whether from farming or another skill, has become increasingly important for women, she said.

The Kibaale District is the third most populous in Uganda, located about 130 miles west of the capital, Kampala.[1] Between 2002 and 2014 Kibaale experienced a 5.54 percent population growth rate, which is much higher than the national average. Nearly half of all girls in Uganda are married before they are eighteen, and 10 percent marry before they are fifteen, even though the minimum legal age is technically eighteen for both boys and girls.[2] Early marriage is deeply ingrained in Uganda's social norms and traditions—along with poverty, a lack of women's rights, and a strong bias against educating girls. Of course, child marriage has many devastating consequences. It not only limits economic opportunities for women and contributes to a higher birthrate, but it also perpetuates the prevalence of poverty and food insecurity.[3]

School, not marriage, provided a future for Safira. And her efforts in school provided a path to prosperity for her whole family. In 2009, when Safira was in her midteens, the family harvested five tons more corn than they could consume, worth $1,000. They used the earnings to start their first savings account.[4]

Safira's journey began when she heard a radio program about a rural development training school in nearby Kagadi. It was private, nonsectarian, and accepting applications for disadvantaged and orphaned girls aged ten to eighteen to enroll free of charge, provided they could pass oral exams in math, English, debate, and drama. One other interesting requirement was a two-generation approach. Parents who enrolled their daughters in school also had to agree to do homework. Their daughter would attend school during the full academic year from late August to early June, and throughout the year the family would take part in one-day workshops in agribusiness, health, entrepreneurship, and apiary management that usually were held the day before she returned home for school break.

In Uganda, and nearly all developing countries, young people bolt from rural areas to seek education and jobs in the cities. Since 1990 the Uganda Rural Development and Training (URDT) program has sought to reverse that trend by providing local men and women with education and vocational training in sustainable livelihoods. Ten years later the program recognized the need to educate rural underprivileged girls and started the school. This school program is unique in linking education to home and community development, thereby allowing girls to flourish in a rural setting.

Safira (and her family) benefited enormously from this life-changing program. She quickly became known as a healer, farmer, leader, and agribusiness entrepreneur in her village. She still lives with her parents and seven siblings. They have become so successful that they have now moved from their eight-hectare (twenty-acre) plot to a thirty-hectare (seventy-five-acre) farm they purchased. But they are no longer subsistence farmers. Rather, they are the top producers of corn, peanuts, and bananas in the district.

For Safira's mother, Mauda Habasa, the family workshops were an effective part of the school experience. "We were enlightened and learned to work together as a team. Our income tremendously improved because of URDT's workshops," she explained. Safira has since received further training as a nurse. She now owns a clinic and pharmacy where she treats infections such as malaria and sexually transmitted diseases, and she sells traditional medicine, prescription medicines, and nutritional supplements.

Empowering girls to transform the livelihoods of the area's poorest families is one of the main goals of URDT, and success stories abound. As a first-year student at URDT, Catherine Namwezi completed an assignment that essentially changed her life. She was ten and in the equivalent of fifth grade when she began an assignment to create a vision of the future: "If all things were possible, what would you like to do?" By writing and drawing her answer, Catherine outlined her vision, not just for herself but also for her family. "We wanted our entire household to have an education," Catherine said. She wanted to be a doctor, a choice largely influenced by her mother, who was a nurse. (Catherine had a brother and sister as well; her father had died when she was ten.)[5]

In the next step of that assignment, which she undertook while home on a school break, Catherine compared her vision to her current reality. The assignment asked:

"What do you have right now?" The family did not have enough income or food. They had to come up with steps they could take to close the gap between their reality and their vision. "My mom and I came up with the idea to grow crops and to build rental houses. We chose crops and rental houses because we were close to the town and wanted to be self-supporting," she said. "If we had this initial money to build rental houses to rent to people, that could be our constant source of income. We built ten of those."

Growing more and better quality food was critical for her family, Catherine recalled. "There wasn't enough sometimes," she said. Before she started at URDT, she typically ate one (occasionally two) starchy meals per day. "We would have tea in the morning and dinner consisting of cassava in the late afternoon. Sometimes there was fruit," she said. "We had poor planning, and at school I learned how to grow food for consumption and plan for the future. We would dry some food to save and eat later. We made lunch and dinner. At URDT they taught us about drought-resistant crops, cassava and banana. [After that] we were never without food." Her family did not have a big piece of land, so they practiced intercropping of cassava (a common starchy food staple), beans, and corn, growing all three together. In her first three years at URDT Catherine learned to support herself and her family through the sale of handicrafts. She made tables, shelves, stools, and bags and taught the skills to other girls in her village when she was home.[6]

Soon after the ten rental houses—built with mud bricks and thatched roofs made from reeds—were finished, Catherine's mom died. Now an orphan at the age of thirteen, Catherine also cared for a younger sister. This left her with little time for crafting, which meant no tuition for her brother and sister and also no pocket money for herself at school. Although her tuition was free, the financial strain made it difficult to stay. At the same time the rental payments from her mom's tenants stopped—she said her renters took advantage of the death of their landlord. "I couldn't get the rent money from them. I was a young girl," Catherine explained.[7]

Still, she fought for her right to an education and improved circumstances. She did not want her dreams for herself and her family to slip away. Eventually, in 2010, an international school in Kampala awarded her a scholarship, and she finished high school in three years, earning an international baccalaureate degree.[8] As this book went to press in early 2017, she was a junior at Hope College in Michigan. Though still quite poor by American standards, she has risen far above anyone's expectations—and her success will profit her whole family. Her brother and sister are old enough now to take care of themselves.

The fight for empowerment and an education lifted Safira and her family from poverty, and it has led her to become a leader in her community. The benefits of education have been manifold. Her family is better off financially. Her siblings have gone to school. She has demonstrated her leadership ability in her village and, most important, improved her family's food security.

Education has also led Catherine to delay the start of

her own family. Family planning is not a direct goal of the URDT girls' school, yet numerous studies have found that remaining in high school is a key determinant of when girls and women begin families.[9] In Mali, women with a high school education or higher bear three children on average; those without an education have an average of seven children.[10] And when a child is born to a woman who can read, that child is 50 percent more likely to survive to its fifth birthday than a child born to an illiterate woman. A thirty-five-year study in Guatemala found a correlation between the years girls spent in school and the timing of childbearing. For each additional year a young woman spent in school, the age at which she had her first child was delayed approximately six to ten months.[11] A single year of primary school has been shown to increase women's wages later in life by 10 to 20 percent, while the benefits of secondary education include a wage increase of 15 percent to 25 percent.[12] Globally, education is nearly a guarantee of improving food security, economic security, and resilience to climate change.

Tying rural education to empowerment and sustainable agriculture is critical for women's development in the future, especially in Africa, where women play a major role as resource managers. Women grow as much as 80 percent of the food crops eaten in Africa—much higher than in the United States, for example, where 86 percent of farmers are male.[13] But women in Africa often have less access to loans, banking, and the best seeds bred to tolerate drought or disease. Safira helped her family plant new kinds of crops, including a new corn that improved yields per acre. The family has used its earnings from surplus grain to buy produce and other nutritious food and to pay school fees for her siblings. Education, empowerment to make choices, and holding title to land have made all the difference between hunger and prosperity in her family.

Safira's story, multiplied by millions of African farmers like her, is exactly why fighting for higher education and innovation in agriculture is so important. Despite being rich in farmland, Africa relies on imports and food aid to feed its people, spending $50 billion a year on food from wealthy countries. Women have fewer rights to landownership, less access to vital technology—such as cell phones—for banking, and less training for improving their agricultural yields. Climate change–related shocks, such as drought, will inevitably impose new burdens on women in Africa. In Uganda only 39 percent of women have rights to titled land, whereas 60 percent of men have rights to land.[14] When women control land rights, they can get loans to improve crop production and manage resources more sustainably.[15]

Getting girls to finish school is only the starting point of this journey, and it remains a challenge. Girls have less access to secondary education, and far more demands on their time, than boys. For instance, in Africa women and girls collect 90 percent of drinking water and firewood, in addition to helping weed and water food crops. As in the cases of Safira and Catherine, educating parents is one of the most effective ways to ensure girls stay in school, in addition to the benefits it offers the parents themselves. For instance, when thirty local women, including Safira and her mother, worked on a land title project, they learned that none of their families owned the title to the land they lived on and cultivated. Because they wanted to invest in better crop production methods, such as irrigation and better seeds and fertilizers, they realized that acquiring title was critical to their production and food security. This quickly led them to establish a regional savings and credit cooperative in their town that lent money to women to acquire land. And the ripple effects continue.

"African countries have previously focused on providing primary education, mostly to its rural populations," Calestous Juma writes in the Uganda daily newspaper *New Vision*. Juma is professor of the Practice of International Development and director of the Science, Technology, and Globalization Project at Harvard's Kennedy School of Government. He also directs the school's Agricultural Innovation in Africa Project, which is funded by the Bill and Melinda Gates Foundation. He explains that Africa's focus on primary education will not prepare citizens for the challenges ahead. In recognition of this, the African Union's draft Agenda 2063 "focuses on consolidating and expanding higher technical training. African agricultural universities have so far tended to training functionaries for the public service. New models are needed to extend higher technical training to women farmers who are the frontline innovators."[16]

The URDT school for girls addresses one of the biggest—

and least discussed—obstacles to sustainability and food security in Africa: fatalism. "People believed they could do nothing to shape their future," Mwalimu Musheshe, chair and cofounder of the school, told a group at the Massachusetts Institute of Technology. "All the outside help in the world would not change this. It only reinforced it."[17] Since 1991 the programs at the URDT campus, which includes a primary school for boys and vocational training for young men, have made it a hub of learning. What was once the poorest district is now thriving. The girls' school, which was started in 2000, now enrolls nearly 400 students annually.

Graduates of the URDT girls' school now have a local option for college, something they never dreamed of. The African Rural University (ARU), an all-women's school, is the most recent addition to URDT's eighty-acre campus. Its mission is to empower women to be effective change agents so that ARU graduates can help the people of Africa improve their lives, transform their communities, and awaken leadership potential in others. The agriculture-based school is the first university in Uganda to offer a course of study in rural transformation and sustainable agriculture exclusively for women. On November 20, 2015, thirty-three women received their diplomas during the university's first graduation ceremony. Twenty-four students earned bachelor of science degrees in technologies for rural transformation, and nine students completed the two-year certificate program in rural entrepreneurship and business management. The graduates are now working as community organizers in seventeen subcounties in the region.[18]

One result of the high number of graduates of the girls' school is the growing number of female high school graduates who meet the ARU's entrance requirements. The university, which has been under development for ten years and recently was accredited, uniquely links agricultural education to community development. The students and graduates develop and deliver programs to village residents of all ages—on topics like creating shared vision, improving health, farming, eliminating corruption, and the empowerment of girls and women. They also develop crucial skills for growing new crops, starting new enterprises, and creating prosperity and health for their families and villages. They learn organic farming methods by working on a thirty-five-acre demonstration farm at the URDT campus, and they participate in broadcasts from a community radio station that reaches 2 million listeners.

African Rural University was modeled on land grant colleges of agriculture in the United States, which were established to extend access to higher education to broad segments of the population, not simply to people pursuing jobs in urban settings. The key to the ongoing success of ARU is the rural focus of the curriculum so that it remains relevant to the lives of women in rural areas.[19]

The rural education focus will likely be key to its ongoing success as well. In a case study, the consultant and researcher Patricia Seybold writes,

> Many universities in third-world countries are located in urban areas and prepare their students for jobs in urban areas. ARU's founders wanted to create an institution of higher learning that would educate people who want to devote their careers to "bottom up," integrated rural development. ARU graduates would work—not in the cities, but in remote villages—to promote sustainable development in rural areas of Uganda and the rest of Africa. ARU education is contextualized in rural communities. One of its goals is to enable national and international experts to listen to, and learn from, rural people in their own communities.[20]

Seybold, who sits on the boards of both URDT and the African Rural University, said the aspiration-based systemic approach is grounded in the understanding that to achieve lasting development, people must become empowered in all areas of their lives, including education, health, economic self-reliance, human rights, and civic participation. "Since its inception, URDT has helped thousands of people improve their lives and has received accolades from international organizations for its innovative approaches. URDT's training of local people, especially women, to become leaders and creators, is changing the way rural communities work," she said in an interview.[21]

ARU graduates now work directly with thirty surrounding communities to help them with their strategic planning and their rural development plans. "It's very hands-on work," Seybold said. A unique aspect of their fieldwork is baseline data collection at each stage of their deployment in the field, she explained. They visit every family. They fill in detailed questionnaires about the livelihood of the family, their quality of life, number of children, quality of health, what crops they are raising, the source of their income, and what issues they are dealing with. The students and graduates return every year and update that information. "They document the rural transformation by collecting huge amounts of fundamental data about the quality of life of these families in these rural communities, and then they work with individual families and communities to identify what each community wants to do," Seybold said. "They document each family's vision for the things they think are going to improve quality of life for the most people."

ARU graduates are employed as rural transformation managers and lead community development as "epicenter managers." Resty Namubiru, thirty-two, was one of the first ARU graduates. She is now an epicenter manager and head of social issues related to gender at the URDT school. When she began her work in rural transformation, much of the challenge, she said, was the physical stamina it required. She and another student would spend a month doing rural fieldwork. At first they gathered population data from households and inquired about residents' interest in agricultural development and participating in local transformation. Resty and her classmate stayed with a host family, but each day they traveled long distances on foot in the hot sun as they sought to cover the ten to twenty villages in the parish.

Her other challenge was communicating her role as rural transformer. In addition to providing informational talks on agriculture and leading workshops to help people envision a different future, epicenter managers guide receptive village households in developing their role in rural transformation. Some villages, for instance, determine that building a road is their priority. Other villages have created savings and loan societies, while still others have created new farmers' co-ops and are working to increase the productivity of farms and gardens.

Resty's main role is to help the community take charge of its own vision. As URDT's Musheshe explained at MIT, "The students need not try to fix the system in place but understand as much of it as they can, and all the forces in play, to constitute the current reality. Once that is understood, they can help community members articulate and use their individual and shared visions to transform that system."[22]

Because ARU is new, one of its challenges is to fill its incoming classes with a diverse student body from other rural areas. Despite its prominence in the Kibaale District, many people outside the region are not aware the school exists, and the pool of potential female applicants is limited to graduates of rural schools, such as URDT, as many young women do not finish their high school diploma. Even if they do graduate, the rural development mission of the school competes with the belief that college should prepare a person for work in a city.

Institutions like the URDT girls' school and ARU are "bright spots" that support the empowerment of women by broadening their worlds, providing hands-on learning, and addressing local needs. In *Switch: How to Change Things When Change Is Hard*, Chip Heath and Dan Heath write, "To pursue bright spots is to ask the question, 'What's working, and how can we do more of it?' Sounds simple, doesn't it? Yet in the real world, this obvious question is almost never asked. Instead, the question we ask is more problem focused: 'What's broken, and how do we fix it?' "[23]

By focusing on women and rural development, the URDT girls' school and ARU instill in their students the core competencies to accomplish meaningful change. The hardest part is identifying the appropriate strategy and resources for expanding their work. By parsing women's education and empowerment goals into realistic projects, and by focusing on community-led solutions, food security will continue to improve.

Safira is now twenty-five years old and remains filled with dreams. "If I had not made a vision, I would not have put up some of what you are seeing," she says in a school promotional video as she gestures around her family's tidy compound surrounded by banana trees and bean plants. Inside the house the pencil drawing that became a road map for her work in progress hangs in a prominent place. "We would go and see [the drawing], and ask ourselves, 'What have we done with our vision?' " she says in the video. "That drawing helped us change our lives today, because we worked towards it." Safira's success, the success of her peers who graduate from ARU, the passion of environmental and agricultural change makers worldwide, and the imperfect choices we have made (and that people around the world will continue to face) in regard to scarce environmental resources are part of a much larger, and equally dramatic, story: the fifty-year global journey of social and environmental sciences, of famine and feminism, and of triumph over tough odds.

Thanks to unorthodox innovators and researchers, and the astonishing results they are producing, the world is at a crossroad. The chapters that follow explore these diverse and productive solutions, many of which were unimaginable just a generation ago when Safira's mother was a teenager.

As Safira's story demonstrates, the more young people achieve food security and a high school–level education, the more likely they are to have an economically and socially sustainable future. How these needs are met—for today's 1.8 billion people aged ten to twenty-four—will define humanity's future.[24] In rural development and agriculture, education provides the skills and knowledge that young adults, especially women, need to be valued and relevant in many developing economies. This is a lever that—if pulled in tandem with instituting policies that ensure land rights for women, the right for women to choose and space the births of their children, and improvements in health care—offers the best hope for feeding a hot, hungry planet.

The (Not So) Hidden Costs of a Lack of Family Planning

The exploding population in sub-Saharan Africa has already had unprecedented and devastating impacts on the environment. Certain threats, such as deforestation, will likely increase in the years to come. As villages grow, and more trees are cut for building materials, cooking fuel, and farmland, more environmental destruction, such as landslides, will occur. In eastern Uganda heavy rains falling in Bududa District, a coffee-growing region in the foothills of Mount Elgon, led to catastrophic landslides that destroyed three villages and killed 388 people in 2010.[25] Three years later massive landslides again occurred after heavy rains, and mud covered five villages.[26] The Bududa foothills are home to a half-million people. With climate change comes more severe rainfall, and when combined with deforested land, it's a recipe for disaster.

Consider Ethiopia, one of the least environmentally resilient countries in the world. Traditional and modern agricultural expansions, continuous land degradation, and urbanization and industrialization have diminished most wetlands in the country.[27] The loss has set off a cascade of consequences, including malnutrition in children; extra workloads for women and the poor, who must travel greater distances to collect water; the absence of traditional medicinal plants; lack of water and forage for livestock; and poor health.

What successful environmental and agriculture researchers, aid workers, and farmers are now beginning to understand is that the answers to feeding places like Uganda and Ethiopia, and many parts of the developing world with growing populations, have been elusive because remedial actions have been taken in isolation. Most agriculture is subsistence farming, practiced in an environmentally damaging fashion that contributes to soil erosion, decreased soil productivity, and water loss. In Ethiopia, rural land is nationalized and is public property. Lack of individual landownership is especially relevant among the extreme poor, who tend to live on smaller or more marginalized plots of land—basically wherever they can stake out a plot. It also has led to severe deforestation, erosion, and desertification. These interconnected challenges are at the heart of recurring food crises in places like Ethiopia. Development dollars have been flowing to specific areas, such as agriculture development, without regard for the effects of that development on the environment, social systems, education, or health care. As a result agricultural development has become unsustainable, and progress has occurred at a cost to the system as a whole.

To help resolve Ethiopia's recurring food crises, projects that integrate complex connections between people and their environment have been introduced in recent years, and they are showing signs of success. In 2013 the United Nations Population Division revised its projections to show that the population would grow even faster than previously anticipated, especially in Africa. By

2014, new projections of population growth again shattered earlier estimates.[28] Development leaders who are planning ahead for sustainably feeding a hot, hungry, teeming planet are sowing complexity into their solutions by dealing with the problem as both a health issue and social issue. One such integrated approach is known as population, health, and environment. The projects have a variety of goals, including poverty reduction, food security, and gender equality.

Without the integration of environmental sustainability, women's rights, and health, many experts remain deeply concerned about sub-Saharan Africa and sounded, frankly, a little panicked when I interviewed them for an article that appeared in *Slate*. "This area is in trouble," said Lewis Ziska, a plant physiologist at the U.S. Department of Agriculture's research service who studies how plants cope with climate change. "We've been screaming about hunger in sub-Saharan Africa for years, not that it has done any good."

Timothy Searchinger, of Princeton University's Woodrow Wilson School and the World Resources Institute, explained, "The problem in crafting a sustainable food future is that it's bigger than people think. Population growth rates are higher and higher. It's harder to keep up with yield growth than we previously thought. The impact of greenhouse gas emissions on agriculture is bigger than people think."

Researchers working on the nexus of food and population are most concerned about the future food needs of sub-

Saharan Africa. Half the world's predicted growth in food consumption comes from population growth, and sub-Saharan Africa is where the population is growing most quickly. "So the question is, what can you do about it?" Searchinger said. "One thing you can do is go around killing people. But that's not going to happen! So we are going to have to find ways to get population back to replacement levels."[29]

That's why the education of girls and empowerment of women are now some of the main solutions for feeding the world, especially because of climate change. Globally, most countries have achieved a birthrate at replacement levels, about two children per woman. But in sub-Saharan Africa the fertility rate is 5.6 children per woman, largely because girls are not aware of reproductive choices, drop out of school at a young age, and have children. In western Uganda, where Safira lives, fertility rates are among the highest in Uganda, at 8.2 children per woman.[30]

Not surprisingly, fertility rates are highest in countries where women don't have access to birth control or maternal and child health care. For instance, 37 percent of women in rural areas of Uganda have an unmet need for family planning. Children often die before they reach school age, and women have many children to improve the chances that some will survive to adulthood. Unless the status of women changes in sub-Saharan Africa, the planet is going to have a lot more than 1 billion hungry people by mid-century.

Robert Engelman, senior fellow at the Worldwatch

Institute and project director of the Family Planning and Environmental Sustainability Assessment, writing in the blog *New Security Beat*, reports that the effect of population on climate change and other environmental changes is so important, "albeit complex and notoriously hard to talk about, that governments are being compelled to explore and acknowledge the relationship." He says that researchers around the world are increasingly recognizing the strength of the population–climate change link. Even the Intergovernmental Panel on Climate Change (IPCC), the official world body for tracking climate change, started including in its assessment how population growth is contributing to the increase in greenhouse gas emissions. Also for the first time, the IPCC reported on the benefits of better access to family planning services.[31]

When women have access to education and are aware of family planning techniques, birthrates decline. The author Gordon Conway explains the connections best in *One Billion Hungry*. He links equal rights for women to improvements in food production and consumption and future advances in food security. Of farmers worldwide, 43 percent are women. Because they are mothers, educators, and innovators, Conway argues, protecting women from discrimination and exploitation, and helping them to be more productive, will prevent widespread famine.[32]

If women are empowered to make their own choices, girls stay in school longer. They have children later in life. If they have access to reproductive health and family planning services when they need them, they can be confident that their babies will grow to be healthy adults, which leads to population growth at replacement rates and improved food security.

People in sub-Saharan Africa are the world's hungriest. In the International Food Policy Research Institute's 2015 index of global hunger, which comprehensively measures and tracks hunger, undernourishment, and malnutrition, sub-Saharan African countries had scores that reflected serious levels of hunger and, as a region, had the highest levels of hunger on the planet. Ethiopia scored twelfth in the world in the hunger index, despite having reduced hunger dramatically since 2000.[33] (In 2011 the average American consumed 3,641 calories per day, while the average person in Somalia consumed 1,695 calories.)[34] But the population growth in the region, along with predictions that people will consume more calories as their countries work to reduce hunger and undernutrition, means sub-Saharan Africa will account for 37 percent of all additional calories required by people worldwide by midcentury.[35] From this vantage point, empowering women may be the only thing that can keep sub-Saharan Africa from starving.

Soils, Sylvan Pastures, and Sustainability

Trees line the gravel driveway that borders the grazing area. At first glance the pasture resembles an overgrown garden on an unkempt tropical estate. Impenetrable shrubby bushes knit themselves together in hedgerows. Grasses reach chest height. Native hardwoods flank the perimeter. But this dense green expanse is not a garden. It is a kind of forested pasture that an increasing number of Latin American farmers are using to raise more cattle on less land while improving the soils.

Until late that morning I was not sure any cattle actually existed at El Hatico, the Colombian ranch I was visiting. Through researchers who were studying the plantings and cattle production of Carlos Hernando Molina, the ranch's owner, I had arranged to talk with him about the transformation of his estate into a sustainable farming operation. El Hatico is located in the rich agricultural valley in western Colombia called the Valle de Cauca. The main house is hundreds of years old, and its walls are two feet thick, which keeps the structure comfortably cool.

Molina's ancestors founded the farm in the eighteenth century. On his watch, however, things were changing. Since the mid-1990s he had transformed 130 hectares (320 acres) of grassland from the conventional barren pasture to fields bursting with special varieties of leguminous trees, shrubs, and grasses. Before we went looking for the cattle, he wanted to have coffee and show me his meticulous records, which tracked production data over several decades.

With this method of grazing he has consistently boosted his income and restored the degraded soil. The plants provide dense layers of food for the cattle, and the vegetation does not need fertilizers (other than manure) to grow. The trees also keep the cattle cool by providing shade. The shade helps to reduce evaporation, and rain that falls to the ground has time to seep into the soil, a process made easier still by the cattle's hooves. As a result his cattle get more nutrition, which has doubled his dairy and meat production per acre while reducing the amount of land needed to raise the cattle. Talk about a win-win.

We urgently need to change the way humans use the global agricultural land base, especially pastures, which comprise the majority of agricultural land. Molina's method is called agroforestry, and it's part of a global trend to sustainably improve agricultural production per acre while reducing the need for chemicals and fertilizers. If farmers, producers, and global growers make these changes, in tandem with the others I will discuss later in the book, the planet has a good chance of meeting the world's food needs after all.

In the pastures near the front of Molina's house are large native tropical trees called *Samanea*, with a canopy shaped like a wide umbrella, as well as mahogany trees. Some mahogany trees are strung with electric wire to form a natural enclosure for the cattle. Inside the forage-filled pen grow five-foot tall *Leucaena* trees, which produce delicate, feathery leaves perfect for nibbling. They are planted in hedgerows on an east-west axis to minimize shade overlap. Three types of shorter plants crop up beneath and between the *Leucaena* rows: African guinea grass, star grass, and peanuts. After a day of grazing, the plants are pruned, and the pen looks as though a weed whacker and lawn mower have attacked it. Molina rotates the cattle between enclosures daily, and each pen is allowed thirty to forty-five days to grow and recover between grazings.

Molina was using a shovel as a walking stick while we toured the fields, and I soon learned why. He pushed his foot into the blade and turned over a clump of soil. The moist clump easily broke into pieces in his hand. It was full of worms and beetles, wriggling proof of the biodiversity benefits of grazing cattle in semiforested pens. While the core benefit of agroforestry is that it allows for more cattle to graze in a given area, it is also recognized as an integrated approach to sustainable land use.

Molina's fields showcase an even more specialized system, known as the intensive silvopastoral system. This type of grazing combines high densities of fodder shrubs, tropical grasses, and trees; it also supports greater biodiversity of birds, mammals, reptiles, and invertebrates. Multiple layers of grasses, shrubs, and forage plants grow together and provide benefits to the land, water systems, and cattle. Trees absorb carbon and add humidity to the canopy, while their root systems help the soil retain moisture. In addition, dung beetles draw animal waste underground, which enriches the soil; in open pasture animal waste frequently remains on the surface, attracting flies.[1] Thus the beetles help to restore ecosystems by enriching the soil with manure, helping to improve the growth of forage. And, because the animals eat and live in better environmental conditions and have access to better-quality forage, they gain weight more quickly, give more milk, and produce less methane.

Conventional cattle ranching around the world simplifies ecosystems by promoting a single grass monoculture as the main food source. The monoculture hampers several key ecological processes, including the rapid decomposition of organic matter, nutrient cycling, regulation of pest populations, and pollination, leading to the loss of species. Intensive silvopastoral systems tend to counter some of the most harmful domino effects of conventional grazing, including soil erosion, which leads to the contamination of water by runoff filled with sediment, manure, and chemicals from fertilizers, herbicides, and insecticides.[2]

There are several silvopastoral systems. Some include the use of small trees whose limbs are cut for fodder, known as the "cut and carry" method. Farmers grow fields of trees and shrubs, such as the nutrient-rich Mexican sunflower, and distribute the fresh cuttings to cows in the forested pastures. But the *Leucaena* is the anchor of the intensive silvopastoral system in this part of Latin America. Its shoots, leaves, and seedpods pack protein. It grows quickly and can withstand constant nibbling by cattle. And its fine, feathery leaves allow sunlight to reach the grass under the plantings. *Leucaena* also fixes nitrogen up to three feet underground, which helps to fertilize the soil naturally.

The Importance of Public Policy

Across Colombia cattle ranchers are making the switch to agroforestry.[3] It is part of an ambitious program to boost farmers' incomes while restoring forests and soil fertility, increasing biodiversity, and slowing climate change.

Livestock and their food production take up 30 percent of the world's land.[4] In Colombia, where the country's 25 million head of cattle occupy 80 percent of agricultural land, conventionally grazed pastures have contributed to severe soil degradation and deforestation, and they have hastened desertification in dry areas.[5] Julián Chará is a researcher at the Cali, Colombia–based Center for Research in Sustainable Systems of Agriculture, a nonprofit that provides technical assistance to farmers who want to adopt intensive silvopastoral systems and that records the progress of farmers such as Molina.

Colombia has strong support for transforming its livestock sector. The country's national cattle ranchers association has set a goal of reducing the area used for cattle production in Colombia by 26 percent, down to 28 million hectares (70 million acres), while increasing cattle numbers from 25 million to more than 40 million head, Chará said.[6] The project, Mainstreaming Sustainable Cattle Ranching in Colombia, had an initial budget of US$42 million, and an additional US$22 million from the United Kingdom's International Climate Fund has extended the project's reach.[7] Molina's work falls within this plan; he offers his fields as test plots for researchers seeking to integrate silvopastoral grazing throughout the cattle-ranching sector.

Agriculture has been the main driver of global deforestation, loss of wetlands, and conversion of grasslands. Worldwide, agriculture has cleared 70 percent of grasslands, 50 percent of savannas, 45 percent of temperate deciduous forests, and 27 percent of tropical forests.[8] Expanding crop and pasture yields is one of the biggest challenges in sustainably meeting the world's future food needs.[9] "To meet projected demands for milk and animal protein, without expanding pasture, production will need to grow by more than 80 percent by 2050," said Timothy Searchinger, research scholar and lecturer at Princeton

University and a senior fellow at the World Resources Institute. "Intensive silvopastoral systems show good promise of getting close to this output in tropical regions."[10]

Fabiola Vega, a Colombian farmer, became aware of the idea of raising livestock with this agroforestry method in 2012 when she heard a presentation at a community meeting about how to raise cattle using sustainable techniques.[11] Her coffee crop had been unprofitable for years. High temperatures, drought, and high costs associated with growing coffee contributed to her decision to raise cattle with the intensive silvopastoral methods. She took most of the land once devoted to growing coffee and devoted it to raising cattle.

Vega's farm, La Cabaña, is located near the town of Armenia in the Andean foothills of Colombia's northern Valle del Cauca. Hillsides are steep. Some hills are still planted with coffee. Others are barren. A 1990s mandate by the Colombian Coffee Federation had called for all coffee plantations to be stripped of trees to intensify the planting of coffee and ensure higher yields.[12] The region had been known for its traditional coffee plantations that had a structural profile of a forest, where the coffee was the understory shrub and fruit trees, banana plants, and hardwoods existed together. As Vega's coffee farm had fewer shade trees and more coffee shrubs, overall coffee production increased but weather extremes and climate change have, in recent years, cut its yields.

To get to Vega's cattle pasture, we walk through a field

of coffee that's now grown in the shade of plantains. When we arrive at her pasture, she walks along the perimeter. "I was hardly getting a return on my coffee investment, so I bought a herd of cows to diversify," Vega explained. A black ponytail streaked with gray falls to her waist. Vega's strength and confidence are at odds with her naturally shy disposition and petite stature. She stands barely five feet tall, and all her life she has worked on a farm. She plants trees and shrubs to keep the barren hillsides from eroding. Forage for the livestock grows where coffee once did, but you won't find the alfalfa or feed corn typical of American farms. Here, the feed is an intensive silvopastoral mix of Mexican sunflower bushes, *Leucaena* trees, and guinea and star grasses. Her cattle have thrived. The landscape now provides environmental benefits such as soil and water conservation and forest cover.

Like any good farmer, Vega keeps an eye on finances. And financial incentives are precisely what helped her transform her farm into a sustainable system. The bank loan she took out to cover the costs of new trees and plants came with a 40 percent discount, thanks to a program backed by the World Bank that has helped 2,200 cattle ranchers convert their grassland to intensive silvopastoral systems.[13]

Vega, who bought La Cabaña in the mid-1990s with coffee profits, will need to pay back only $12,000 of the $20,000 loan she received to improve her pastures. More important, her farm is now economically stable. The calving rate of her cows was 100 percent in 2014, her first full year of operation. Compare that to the average calving rate for conventional cattle ranches in Colombia—52 percent.

Perhaps most important, the new farming methods provide Vega with greater resilience to climate change—both economically and socially—and improved food security. With the income derived from the farm, she has hired additional staff and persuaded her son and other family members to return to the farm.

As cows produce more milk, they also need to be milked more often to take full advantage. Molina can have his cows milked twice a day because his farm has storage and is near a metropolitan center. Vega's cattle are milked just once a day. Because her farm is on the northern edge of the Valle del Cauca in the former coffee region, it is too far from the dairy infrastructure for a twice-a-day collection service. With new improvements to infrastructure and loan incentives, Vega hopes to soon have two milkings a day.

Farmers usually achieve a 100 percent return on investment because of increases in milk production if the cattle are milked twice a day and cattle gain weight at a faster rate, according to an analysis by Enrique Murgueitio, a researcher and executive director at the Center for Research in Sustainable Systems of Agriculture. Within four months of establishing plants and fodder shrubs in her fields, Estella Dominguez said her cows could be milked twice a day for the first time. And she achieved more financial gains because she didn't have to pay for fertilizer or herbicides on grazing land.[14] Like Dominguez, most farmers demonstrate a return on their investment in two years.

About a half-hour drive from Vega's farm is Pinzacuá, a forty-five-hectare (110-acre) ranch owned by Olimpo Montes, who has owned it since 1983. While the farm is now a densely forested cattle ranch, it didn't always look so lush. On a wall in his living room hangs a satellite image of the farm in 2003—an expanse of desolate hillsides. But when a coffee crisis that began in the late 1990s saw Montes's operation running in the red, he decided to reforest the property and change his operation to an intensive silvopastoral grazing system for cattle. Like Vega and Molina, he now grows grasses and protein-packed trees.

Montes made the switch to agroforestry almost overnight. He jokes, "I went to bed wanting to be a coffee producer and woke up a cattle rancher." Coffee had failed to provide him a profit, he said. His property suffered from landslides and soil erosion. So, beginning in 2002, with the help of a Payment of Environmental Services—compensation he received following changes he made as a participant in a project funded by the Global Environment Facility and administered by the World Bank—he transformed his formerly treeless land into pastures filled with trees. He started with one hundred head of cattle and used all forty-five hectares (110 acres); now he raises a herd the same size on twenty-five hectares (60 acres). On the rest of his land he produces shade-grown coffee, plantains, pepper, vanilla, and high-value hardwood timber.[15]

Vega's, Molina's, and Montes's brand of sustainable intensification is the favored agroforestry practice for livestock production in Colombia. Molina has trained cattle producers from other parts of Colombia, as well as Mexico and Costa Rica. Additionally, producers from Nicaragua, Panama, Brazil, and Rwanda (and more developed countries like New Zealand and Australia) have visited the Colombian farms that are using intensive silvopastoral practices; the producers want to increase their own climate resiliency and livestock production while conserving land.

Murgueitio is the man behind the development of this brand of sustainable cattle ranching in Colombia. He holds a doctoral degree in veterinary medicine and animal husbandry. Farmers who visit his center's test locations, including Molina's El Hatico, learn how to replace conventional pasture stubble with agroforestry methods that use 4,000 plants per acre—a mosaic of trees, forage shrubs, tropical grasses, and groundcover—to feed their cows. While countries like Brazil are now ramping up their cattle production using industrial agricultural systems like feedlots, Murgueitio takes the long view. He argues that a simpler, more sustainable, approach makes better economic and environmental sense.

Most of the cattle that thrive in Colombia are a mix of Brahmin and South American cattle. They have a long history of withstanding the heat and pests of tropical regions, but finding out how to feed them a healthy mixture of trees and shrubs has been Murgueitio's primary job. Standing at Molina's ranch amid low-slung forage trees and shrubs,

Murgueitio takes in a deep breath. The tropical sun beats hard on the grassland a few steps away, but it is a noticeable five to ten degrees cooler in the light shade of the silvopastoral grazing area. The soil beneath his feet is rich and fertile, an important feature of grazing at El Hatico, where the roots of this system go back more than twenty years.

"The advantage to this system is that the benefits to the environment and to the rancher occur simultaneously," Murgueitio said. "We are building systems that can be applied to traditional ways of production that not only take into account the value of the livestock production but also justice, equity, equality, democracy, solidarity, and being respectful with nature. A silvopastoral system is sustainable, and in that way is respectful to the next generation."[16] Sustainable cattle ranching and financial gains go hand in hand.

Evidence-Based Agriculture

The simplicity of the silvopastoral system, however, doesn't mean it is not backed by strong science. Just the opposite. People are turning to solutions like silvopastoral systems precisely because the science is so clear, Murgueitio said. Extreme weather events, such as those associated with La Niña in 2010 and 2011, led to yield losses across the agricultural sector, but producers using silvopastoral systems were better able to cope with the changes.

However, as with many sustainable solutions, high

up-front costs have posed a barrier. Farmers need expertise to plant and maintain the trees. Historically, cattle ranching was a simple operation. Aside from fertilizers and herbicides to help the grass grow, conventional cattle-grazing operations require few resources, according to Chará, the sustainable agriculture researcher. Historically, ranchers don't like large trees that cast heavy shade because cattle tend to sleep—not graze—in the shade. That leads to a muddy, grassless expanse beneath the tree. But both images run counter to the realities of intensive silvopastoral systems, where the trees have small leaves and provide cattle with light amounts of much-needed shade to escape the tropical sun, so they suffer less heat stress. And because the forage is thick beneath the shade, they also can graze throughout the sunniest parts of the day and thus produce more milk. But bridging knowledge barriers, and advocating for something that runs against the accepted wisdom, is always difficult.

As a result of policy incentives from the Colombian government and support from the country's national cattlemen's association, along with the free and easily available technical assistance from the Center for Research in Sustainable Systems of Agriculture, farmers from Mexico to Brazil are now using Murgueitio's sustainable system. By 2015 sixteen of Mexico's thirty-one states had networks of intensive silvopastoral grazing systems, with 10,000 hectares (25,000 acres) planted.

The tropics have the unique ability to both support the growth of biomass for cattle feed and increase the amount of carbon held in the landscape, Searchinger told me. "Increasing the number of trees in pastures on a large scale in a scientific way, as a method of increasing production while absorbing carbon, is new," he said. "The question now is, how can you expand it at a really rapid rate? With intensive silvopastoral systems, we've identified the magic bullet, but we don't yet know how big this bullet must be to solve the problem." Production under intensive silvopastoral systems can be 70 percent higher than otherwise well-managed and fertilized pastures, he said. "Grazing produces half the world's animal feed. That's a big deal," he said. "If everybody would do it—if every livestock producer in Latin America shifted to silvopastoral systems—it would solve a lot of land-use problems."[17]

Although scientists have demonstrated that intensive silvopastoral systems have benefits for forest ecology and natural systems, can increase a region's resilience to climate change, and improve a farmer's income, economic and policy barriers can be major hurdles. Most landownership in Colombia is informal, Searchinger said. Because smallholders do not register titles, banks cannot grant them loans for planting trees and improving pastures. Few alternatives are available for financing agroforestry on informally held land. And, with Colombia's long history of violent land takings, the problem of landownership will not likely resolve itself soon.[18]

Some people may frown at doing anything to make cattle production more sustainable, because emissions from ruminants are one of the largest sources of greenhouse gases. But livestock provide important nutrition for much of the world, so failing to try the most environmentally efficient approach possible would be shortsighted. “Cattle ranching is seen as ecobad no matter what,” said Amy Lerner, a researcher at Princeton University who has been studying agroforestry in Colombia. “But there is a huge portion of global land area already in pasture. Is there a way we can make these pastures better?”[19]

In fact, interventions in the pastures can reduce emissions from livestock while also improving their production efficiency. Because cattle digest this forage more easily, they produce 20 percent less methane. In addition, silvopastoral systems increase carbon sequestration in both trees and soils and reduce the need to use fire for pasture management.[20] Intensive grazing may seem an unlikely partner in sustainable land management; however, proven benefits for rural economics and food security, and even more so for environmental services and reforestation, might help ranchers understand the advantages.

Paul West and other researchers at the University of Minnesota have identified regions, crops, and actions that can increase food yields, reduce the environmental impact of agriculture, and more efficiently deliver the food grown. One of the goals of their research is to grow more food on existing land in ways that limit the stress agriculture puts on natural ecosystems. His analysis shows that prioritizing a small set of leverage points could greatly increase the efficiency and sustainability of food production.[21]

In addressing social problems, the effort to prepare for

change often dwarfs the actual making of the change. For example, the civil rights movement took decades of demonstrations—often by people who put their lives on the line—rallies, speeches, personal conversations, and lobbying for new laws to prepare the country to make the change. To actually change the system required passing new laws and then enforcing them. Likewise, lead in gasoline had been a public health concern since 1925. That didn't change until the 1970s, when public concern about air pollution prompted legislation such as the Clean Air Act and eventually the requirement that catalytic converters be installed in vehicles to remove toxic pollutants (the converters work only in vehicles that use unleaded gas). Similarly, solving the world's food production problems in the next three decades in the midst of climate change will require changes and innovations and different choices. The farmers in this chapter demonstrate how changes on the ground translate into more sustainable systems. Molina changed his farm landscape, leading to a cascade of beneficial effects for the cattle, soil, water, and ecosystems. He has even attracted 135 of the region's 141 bird species to his farm, 66 percent of which were counted on land he had converted to his silvopastoral system.[22] Murguietio innovated change by methodically and systematically researching and developing science and animal husbandry to support silvopastoral grazing for healthier livestock, healthier soils, and healthier ecosystems for beef and dairy for farmers. Vega took a risk, switching from coffee growing to a cattle operation and learned a new way of life. Put together, these farmers and researchers demonstrate how improving the way humans produce more food from the same amount of agricultural land can improve food security. The changes and choices Colombian farmers have made, supported by a well-coordinated national plan and smart policies, economics, and ideals of sustainability, will help build a critical mass capable of bringing about a national change in Colombia's agricultural systems. As farmers in other countries, including Rwanda, Brazil, Mexico, and Nicaragua, among many others, visit Molina and develop their own version of Colombia's intensive silvopastoral system, they will help resolve the issue of sustainably feeding more people on less land.

California and Syria

A Tale of Two Droughts

In the late spring of 2014, while covering food sustainability at the Monterey Bay Aquarium's Sustainable Foods Institute, I took a trip to the Carmel Valley farm stand run by Earthbound Farm. Earthbound Farm is the largest organic farming operation in the United States. It cultivates about 50,000 acres of produce, and I spent the morning walking in a small demonstration garden that was nothing short of paradise. Everything was a verdant green. Yet just beyond the farm, where the Carmel Mountains meet the horizon, was dry scrub and pale brown grass, a truer reflection of this parched land. The Golden State, which got its name from the grasses that turn a shade of palomino blond in summer, then green up again during the fall and spring rains, was looking more like the Brown State.

As California's drought dragged into the next year, I couldn't shake the sense of a crisis brewing in Carmel Valley. I was also hearing reports of conflict over water in war-torn Syria. I wondered, could water conflict on that scale ever occur here? I couldn't blame Earthbound's owners for choosing this idyllic spot, or other farmers for choosing any other location along California's central coast, where morning fog moistens the otherwise dry landscape. When the founders of the farm first started growing raspberries on two and a half acres, they didn't imagine it would expand to become America's largest organic producer of salad greens and vegetables. But Earthbound's growth was only one among the more recent in decades of farming expansion all across California, and especially the nearby Central Valley, since the Dust Bowl of the 1930s. Through the magic of irrigation, these farmers had made a desert bloom.

While Earthbound's leafy expanse appeared intact, agriculture is in jeopardy throughout California and other western states. A 2015 investigation in *ProPublica* reported that California's drought is part of a much bigger water crisis that is killing the Colorado River, "the victim of legally sanctioned overuse, the relentless forces of urban growth, willful ignorance among policymakers, and a misplaced confidence in human ingenuity."[1] Climate change will only exacerbate the problem.

The state's farmers are managing to stay in business by using dwindling water resources, but it's a heavy toll. The record-setting drought began in 2012 and now includes, among other indicators, the driest calendar year and the highest annual temperature, leading to acute water shortages, overdraft of groundwater, critically low flows of water in streams and rivers, and enhanced risk of wildfire. A report in August 2015 predicted that the extreme water shortage would cost California's agricultural sector $184 billion and more than 10,000 workers their jobs by year's end.[2]

California's farms rely heavily on groundwater. Pumping water from deep underground during this severe drought has limited farmers' crop losses, but it's ultimately a losing battle. Pumping from deeper in the aquifer increases farmers' energy costs. And when those wells run dry, they hire contractors to drill more wells. Some basins in California's Central Valley are being so severely overpumped that the aquifers can't replenish themselves. This invites a host of other problems, including sinking land (the gradual settling of the Earth's surface, up to two inches per month in some places), water quality problems, and diminishing reserves for future droughts. It's been a wake-up call for policy makers who fear the reserves have gone far too low.[3] Scientists at the National Oceanic and Atmospheric Administration say California's water supply will need several years of above-average rainfall before it can return to anything close to normal.[4]

California and Syria are similar in their Mediterranean climates and experience of drought, which makes them ripe for comparison with regard to water use. Two recent studies that focus on severe droughts in each location illustrate the widening gap between the abilities of established economies like California's and those of a developing country like Syria to adapt to the effects of climate change.[5]

In fact, Syria is a textbook illustration of how dwindling water resources in developing countries shock food security, prompt conflict, and disrupt national security. Take, for instance, the well diggers in northern Syria, who worked successfully for decades. Farmers with cash hired well diggers to ram pipes 200 feet underground to reach aquifers that would ultimately irrigate the farmers' fields. Using heavy but reliable equipment, and day laborers eager to do the grunt work, the well diggers knew where to dig and knew which government officials would ignore them if they bent the rules, according to *Scientific American*. Then came the drought. From 2006 to 2010 water levels plummeted. Pumping continued unabated. When their wells ran dry, farmers dug new ones. Although the government required that farmers apply for and receive a license to dig new wells, the practice was overlooked for a price.[6]

During the first part of the drought, in the winter of 2006–2007, the water table fell like never before. The well diggers suddenly had to dig 300 to 600 feet underground before hitting water. During the worst of it, diggers thrust pipes nearly a half-mile deep.[7] Adding illegal wells to an already depleted aquifer meant that water drained disastrously quickly. Colin Kelly, the lead author of a study published in the *Proceedings of the National Academies of*

Science and a PACE (postdoctorals applying climate expertise) fellow at the University of California, Santa Barbara, explains, "What's happening globally—and particularly in the Middle East—is that groundwater is going down at an alarming rate. It's almost as if we're driving as fast as we can toward a cliff."[8]

Agriculture and water specialists often cite California's water woes in the same sentence as Syria's. Water and climatic conditions played a direct role in the deterioration of Syria's economy and society.[9] When work ran out in the rural areas, farmers migrated to Syria's cities, adding to the melee. These migrations occurred as Iraq War refugees were streaming into Syria. Roughly 1.5 million farmers and rural people fled the countryside for the urban outskirts, where social unrest boiled over into civil uprising. Syria's urban population rose from 8.9 million in 2002 to 13.8 million in 2010. Kelly's study summarizes it this way: "The rapidly growing urban peripheries of Syria, marked by illegal settlements, overcrowding, poor infrastructure, unemployment and crime, were neglected by the Assad government and became the heart of the developing unrest."[10] Environmental security—in the form of stable water resources—affected food security, which cascaded into a violent meltdown.

The drought was one factor in the loss of water supplies. But lack of strong policies to prevent overpumping was also to blame. Eventually, the money ran out, the water ran out, and people abandoned their land. The rich topsoil—known in the Middle East to provide some of the best yields of wheat, fava beans, tomatoes, and potatoes—had become desert-like and lifeless.

Syria is one of the world's longest-inhabited regions. The moist river delta region known as the Fertile Crescent—which has supported agrarian civilizations since 12000 B.C.—is drying out, hastened by climate change.[11] In modern times agriculture continues to be one of the main industries in Syria. The unstable crop production, largely dependent on rainfall and irrigation from increasingly diminished wells, eventually led to failed harvests and abandoned farms. In turn, global prices for major grains shot up and helped to trigger outbreaks of civil unrest in more than forty countries. For instance, when cash-strapped governments reduced subsidies for bread, a major staple, the decision helped spark uprisings such as the Arab Spring. (In 2010 Egypt spent $3 million on these bread subsidies. Analysts suggest investing that money in agriculture would have been more effective.)[12]

The food insecurities that arose following Syria's extended drought are a sign of decreasing resilience to environmental shocks to the global food system, experts say. After World War II mechanized agriculture, irrigation, fertilizers, new varieties of plant species, and innovations of the green revolution led to increases in crop yields. Most famines in recent decades have been the result of limited access to food, rather than lack of food itself, and policies that support access to resources have greatly influenced food security (for instance, the policies that approve or deny permits for well digging). The world population is also putting so much stress on the global food system that it's challenging its resilience. The global population has increased by 60 percent since the 1980s, while crop yields have plateaued.[13]

Changing environmental conditions will likely make lack of access to food a regular occurrence. What does this imply for the global food system? An alarming analysis, summarized in a 2015 study by Samir Suweis and colleagues, reports in the *Proceedings of the National Academy of Sciences*:

> The emergence of global food scarcity is confirmed by a number of recent studies suggesting that the limited resources of the planet [primarily land and water] would soon become insufficient to meet the escalating demand for food, fibers, and biofuels by the increasingly numerous and affluent human population. Such conclusions indicate that we may be at the verge of a global-scale Malthusian catastrophe. One then wonders how sensitive global food security is to perturbations arising from drought occurrences, changes in energy and trade policies, or food price spikes.[14]

According to this study, the global food system is losing resilience, becoming increasingly susceptible to crises such as drought, floods, extreme heat, and wildfires. The instances of food shortages and peaking prices in Syria and elsewhere in the Middle East between 2006 and 2011 are consistent with the findings of many researchers. The pressure is on agriculture globally because of the ever-increasing demand for agricultural products as a result of population growth, shifts to more resource-demanding diets like beef and chicken, and new bioenergy policies that promote the conversion of corn and other crops to biofuels. This is happening while crop yields have been stagnating. "Humanity is rapidly exhausting the safety margins underlying the resilience of the global food system," Suweis and colleagues write.[15]

The global food trade allows populations to consume food produced elsewhere—and it has become crucial to the world's food security. Egypt, for example, cannot easily grow the full range of foods its citizens require. The country has become the world's largest importer of wheat, bringing in 11 million tons annually. Wheat is used to make flat bread, the Egyptian staple. Egypt grows a lot of wheat itself, producing roughly 9 million tons a year, but the country needs the imported wheat to mix with domestic wheat to increase its gluten content to make it suitable for baking bread.[16] Egypt also imports an increasingly large amount of livestock and dairy products from developed countries such as the United States. Indonesia, too, has experienced a surge in agriculture trade. Its agriculture imports quadrupled between 2005 and 2015, to $18.4 billion, even though its economy is largely based on agriculture. Now Indonesia is the eighth-largest trading market for the

United States and the largest market in South Asia for U.S. agricultural exports.[17]

But the United States isn't the only supplier. A growing number of trade partners have entered the global market since the mid-1980s. It may not be a surprise that China is the leading producer of rice in the world. But it is also the largest producer of wheat and the second-largest grower of corn.[18]

So what, exactly, is it that links food security to the economic security and political stability of a country? Emmy Simmons is trying to find out. After a thirty-year career at the U.S. Agency for International Development (USAID), the last three years as assistant administrator, she is working with AGree, an organization that advises on international food and agriculture policy. In this role she helps domestic and international leaders bridge the divide between what the United States does in food and agriculture and what the rest of the world does in food and agriculture. We met for several hours in the summer and fall of 2014, and she shared some of what she's found in the tangle of natural resource management, food security, and global conflict.

One point that stood out for me was how disconnected American farmers are from the rest of the world. Simmons meets with farmers as part of her job at AGree, supporting her group's mission to drive positive change in the food and agriculture system. She has consistently found a gap in knowledge. International food aid organizations that employ mostly Americans have little understanding of what is important to American farmers, and the converse is often true: American farmers generally have little understanding of global farming.

"On one side are Americans who have a familiarity with global policy concepts and who pay attention to emerging international movements such as the Millennium Development Goals and the Sustainable Development Goals. They have a certain familiarity with a kind of long-term history of food policy with India, South Korea, and Australia, which has done some amazing work on food and agriculture," she told me, explaining that Americans who understand the intricacies of global food politics don't know much about the decisions and politics important to the U.S. farmer. "There is this gigantic divide [in understanding] and one of the things that I've been struggling to comprehend is that what U.S.-based thinkers think domestically and do domestically, and what's happening in the rest of the world, are separate dialogues."[19]

An issue fresh in her mind was the rise in corn and soybean prices in 2012 as a result of the long-term drought in the American Midwest. In "Harvesting Peace," a report for USAID and the Wilson Center, Simmons writes,

> The possibility that increasing prices would translate into a new round of increased worldwide food insecurity in 2013 was worrying news. The negative political and human security effects of the food riots that began in 2008 were fresh in leaders' minds. The eruption of violent conflicts in rural areas of Mali, Sudan, and

South Sudan continued to link the issues of food and conflict in the media headlines.[20]

After decades of inexpensive and abundant food, ever-increasing varieties, and year-round availability, most countries were unprepared for the political and economic shockwaves that ricocheted around the world between 2008 and 2010. Food availability is just one aspect of food security; another aspect is the ability to get food at an affordable price. Between 2005 and 2008 food prices rose on average by 80 percent, according to the International Commodity Price Database of the Food and Agriculture Organization of the United Nations (FAO). The world was gradually consuming more of its grain stocks each year. As demand rose, so did prices. For example, the world price for rice rose by 217 percent between 2006 and 2008, and the price for wheat by 136 percent. American households did not feel the rise as much as households in other countries, because the cost of basic grains like rice, wheat, and corn is not a large proportion of an American grocery bill. However, skyrocketing prices sparked riots from Mexico to Pakistan. Reports of public unrest over food prices also came from Haiti, Malaysia, Indonesia, the Philippines, Bangladesh, India, Burkina Faso, Senegal, Cameroon, Morocco, Mauritania, Somalia, Ethiopia, Madagascar, Kenya, Egypt, Ivory Coast, Yemen, the United Arab Emirates, Mexico, and Zimbabwe.[21] According to the FAO, thirty-seven countries faced food crises, attributed to increasing demand from the growing middle classes in

China and India, extreme weather events that had reduced harvests, a spike in energy prices, and the growth in production of farm biofuels.

On the supply side, rapid increases in the cost of energy were partly to blame for price increases. Energy prices, especially the price of crude oil, are closely linked to prices for fertilizer, a petroleum-based product. Energy costs also drove up the cost of transporting food. On the demand side a number of factors contributed to reductions in grain stocks, including the diversion of grains to biofuels and the income and population growth that led to changing dietary preferences. Technically, farmers today grow enough food to feed everyone. But nearly 800 million people on the planet periodically go without eating, in most cases because the food is too expensive or not available in the right places. That number could get a lot bigger.

🌐

Simmons has tried to educate her colleagues at U.S. development and aid agencies about the connections of global food security and management of water resources. As she talks with them about subjects like groundwater depletion in California and the future of water distribution from the Colorado River Basin, she finds that her colleagues in international development are unaware of the issues and the volatility of water resource claims in the western United States. They did not understand how California's reliance on irrigated agriculture had much in common with the lack of water in Syria. If Syria's war started with water, should

California, the world's seventh largest economy, expect economic instability that could ripple through the rest of the nation? Syria's agribusinesses overpumped groundwater, which accelerated migration from rural areas to cities and increased tensions. What tensions might occur if California agribusiness is allowed to overpump, and who will be responsible for the cascade of effects? Drought intensified mass displacement in Syria. In parts of California, wells ran dry in 2014 with no long-term solution in place. Will residents without water need to pack up and leave?

Similarly, from discussions with U.S. farmers, she has learned that a farmer's impression of America's role in agriculture did not match reality. "One of the things that gets farmers up to go to work every day, they tell me, is that they are feeding the world," she explained. "They say this with a straight face, with lots of emotional content. They say, 'We American farmers are feeding the world.'" Simmons responds by explaining the high yields that other places, such as Brazil, China, and India, are now achieving. "I say to American farmers, 'Of course you know that you're not! You're not feeding the world.'" The farmers disagree passionately.

Other reports about farmers' opinions mirror Simmons's experience. In 2013, at a farm convention in Illinois, the seed and chemical company DuPont created an interactive wall that invited farmers to write their answer to the question "How are you making a difference to feed the world?" An Indiana farmer who grows corn and soybeans on thousands of acres explained to a reporter that he was aware of the growing demand for food from the burgeoning global population. Then he cited the decline in the number of American farmers. He told National Public Radio, "It's the duty of those of us who are left in the business, us family farmers, to help feed that world." The farmer explained that he also understood that preferences for foods from higher up the food chain are increasing. He cited global pressure to feed the world as the reason for growing more food per acre, a process that industrial agriculture production and technology, such as genetically engineered seeds and pesticides, strives to aid.[22]

Such explanations grate on Simmons. She says the moral argument for increasing the use of pesticides, irrigation, production, and improved seeds to "feed the world" neglects the big picture: the true amount of corn, soybean, and wheat the United States exports and what it means in a global context. Simmons says that when she hears farmers say, "Other people can't live without us!" she responds with a question: Do you know what percentage of global production you contribute? The farmers consistently overestimate their contributions to the global food system by three times, Simmons said. Also, half of all corn grown in places like Iowa goes to producing ethanol, which is added to gasoline—that corn isn't feeding anyone. In truth, American farmers account for 12 to 14 percent of global food production. Their share of world corn and soybean production is declining as other countries boost production. By 2014 corn accounted for only 6 percent of U.S. agricultural export value, down from 12 percent in 2012.[23] Although the United States is the world's largest corn exporter by volume, exports account for a relatively small share of the overall U.S. corn crop.

If we allow Big Agriculture to argue that farms must increase their use of fertilizers, chemicals, and water in order to feed the world, we are guilty of hastening the demise of sustainable American farming. This is the same argument agribusiness used to suck the Colorado River dry. Why do we deplete the Colorado River and groundwater in California? Why apply pesticides that damage the environment and provide mediocre food? The reasons are ultimately far more self-serving than altruistic. Let's take a closer look at California.

Case of California

If you take 80 percent of something people value, scrutiny and criticism will probably follow. That's what California farmers who irrigate are learning, because agriculture uses nearly 80 percent of the state's water. In turn, the state's growers supply a third of the vegetables in the United States and two-thirds of the fruit and nuts. California is the largest producer of almonds and pistachios in the world.[24] Enjoy a domestic artichoke, and it almost certainly came from California: the state supplies nearly 100 percent of the U.S. artichoke crop, four-fifths of it from just one county,

Monterey, where the town of Castroville calls itself "the Artichoke Capital of the World." (A number of Mediterranean countries produce more, and no doubt reject any U.S. claim to artichoke supremacy.)

In 2015, after emergency drought legislation forced cities and towns to cut water use for the first time (after they had endured a spate of vitriolic barbs about "watering lawns in the desert"), California's irrigated agriculture became a prime target for criticism.[25] And the almond is taking most of the fire. Producing one almond requires about a gallon of water, and almonds account for 11 percent of the state's agricultural water use. Fruit, nuts, and alfalfa combined require 40 percent of California's freshwater withdrawals, and they, too, are getting a critical look.

Josué Medellin-Azuara, a senior researcher at the University of California's Davis Center for Watershed Sciences, said that fruits and nuts are grown on about one-third of the state's irrigated cropland and use a third of the water, but they account for almost half of all state crop revenues. Those figures compound the challenge of an already troubling water future. Add to it that permanent crops and irrigated crops like alfalfa and feed corn provide additional value along their supply chains, and the horizon only gets darker.

Water irrigates two categories of crops in California: crops that are planted seasonally in fields, and high-value permanent crops, such as orchards, which grow for decades. Since the 1990s the area of land planted with permanent crops has significantly increased. Fruit and nut trees and grapevines need a minimum amount of water year in and year out, and they cannot lie fallow. Seasonal crops have flexibility when it comes to water availability—and fields can go unplanted during drought. The amount of land under permanent irrigation also is increasing; since the mid-1990s many farmers across the state have switched from cotton and alfalfa to almonds. "These are high-value crops and much higher revenue per unit area, which makes these attractive for farmers," Medellin-Azuara said. "But if we keep increasing permanent crops in the state, it reduces the flexibility in the system."[26]

According to the Public Policy Institute of California, the shift to high-value perennial crops has increased the value of the state's agricultural production from $16.3 billion in 1998 to $36.4 billion in revenues in 2012.[27]

"Having too many [permanent crops] reduces the way we cope with drought and makes fallowing more expensive," Medellin-Azuara said. Yet flexibility in agriculture is just what the state needs in a future characterized by a warming climate, said John Matthews, a water ecologist with the Alliance for Global Water Adaptation. Fruit and nut trees are a lot like having a big, rigid infrastructure that requires vast quantities of water for decades, he told me: "That is a serious commitment, and it sets a clear floor to your water commitment for a long time."[28]

With an abundance of high-value crops that need a fixed amount of water, California will find it difficult to modify its agriculture economy to accommodate a more drought-prone future, according to Matthews. He said the ricochet effect of overpumping groundwater, which takes decades to recharge, is like watching a slow-motion train wreck. Because of climate change, extreme weather conditions and severe drought will occur more often than they have since California's economy was built. Globally, agriculture uses about 70 percent of water withdrawals, so California's 80 percent share isn't that much higher than the norm. But it does make the state more vulnerable to extremes.

California is now attempting to plan for this uncertain future, but the water warnings are more than a decade old. A report published in 2004 in the *Proceedings of the National Academy of Sciences* was just one of several studies suggesting problems ahead. "Flexibility really is the key to climate preparedness," says Katharine Hayhoe, lead author of that study and a climate scientist at Texas Tech University.

> The main reason why we humans even care about climate change is because we have built our vulnerabilities into our systems. A thousand years ago, if climate shifted or sea level rose, what would we do? We'd pick up our tents and our farming implements and move. Today, that option is no longer available. That's why we need to be more flexible in the places where we live, in many ways; from building new neighborhoods that float, in the Netherlands; to plowing biochar [charcoal from plant matter] back into our fields in Iowa to better retain water and nutrients as well as sequestering car-

> bon; to diversifying our crops such that they are able to withstand very different future conditions than the ones we have seen in the past.[29]

Fundamentally, agriculture needs water, and certain crops need far more than others. Drought offers clear challenges but also an opportunity to diversify, to seek out better management practices, and to improve efficiencies. A key question, as yet unanswered, is whether water use for agriculture in crop-rich California can coexist with the other needs of the state's growing population. And, if so, how?

The problems in Syria and the drought dilemma closer to home could lead to the adoption of better methods and incentives to conserve and enhance the natural resources agriculture depends on, as Hayhoe suggests. Agriculture is often at odds with environmental protection. But it need not be an either-or equation. One big challenge for the United States is to shift its expertise. The United States has enormous capacity to improve plant cultivars and livestock breeds. This same capacity could lead to better management of the natural resources needed in agriculture. With more integrated approaches, agriculture could become a pathway to addressing the scarcity of natural resources by making sustainable production, conservation, and resilience complementary techniques rather than notions that elicit antagonism.

Decision makers in many areas could benefit from knowledge of how the global food system works and the precarious position of the food, water, and energy nexus. Water is being overexploited and polluted like never before, and agriculture is a major cause. As a result of the overexploitation, interest in western water problems is increasing. People are beginning to question why we are irrigating alfalfa in California so we can export hay to China, Japan, or the United Arab Emirates. Why are we doing that? Why are we taking water—one of our most precious resources, especially in the western sector—and exporting that water abroad through crops? The water contained in agricultural products, like corn, soybeans, wheat, beef, and chicken, is called virtual water. When water-intensive commodity food grown in California or Iowa is shipped to places like Japan or China, those countries play a direct role in the hydrological cycle of the United States. As *National Geographic* reports, Robert Glennon, a water policy expert at the University of Arizona, estimated that in 2012 roughly 50 billion gallons of western water, which is enough to supply the annual household needs of half a million families in the United States—were exported to China, embedded in the alfalfa crops sent there.[30]

Understanding how much water is going to what areas of the world is critical to developing agricultural efficiency. There's a direct link between Japan's or China's demand for water-intensive crops and the water used for the production of commodity food grown in the United States. This water used for exports contributes to the change in regional water systems. The almond trade in California is one example, with its use of water from the Colorado River for irrigation. Governments and policy makers are now taking a greater interest in whether they can purposefully interfere in the water politics and allocations of the western states to achieve increased efficiency.

We can never know everything about sustainability in the global food system, especially given climatic variability. Sustainable agriculture is defined in simple terms as the production of food, fiber, plant, or animal products using farming techniques that protect the environment, public health, communities, and animal welfare.[31] With each new pressure on limited resources, whether water, land, or soils, discerning which is the sustainable path will become more difficult.

The people working in the field of sustainable agriculture are now the metaphorical rafting guides on the journey along the turbulent river, Emmy Simmons explained. As additional streams enter a river and cause waves from water turbulence, the river often becomes more complicated as its journey continues. That's exactly what those working in agriculture research are seeing now. Fluctuations in rainfall and temperature are unlike what has occurred in the past. Pressures on water, land, and forests from a growing world population that is eating more water-intensive global commodity foods have the metaphorical rafting guides unsure of how to navigate the journey.

As the world becomes increasingly dependent on the global food trade, we are losing our ability to cope with shocks and stress.[32] To understand the magnitude of the many solutions required, and which ones are even possible, it is necessary to first understand how we became so dependent in the first place. That is the subject of the next chapter.

hungry

Population, Health, and Environment Powerfully Working Together

Until well into the afternoon, the crowd roared louder and louder, and fists pumped in the air. Freddie Mercury strutted up and down the stage in a white tank top and faded jeans, hitting all the sustained high notes in "Radio Ga Ga!" The 72,000 fans at Wembley Stadium in London roared with his refrains. A banner stretched above his head read, FEED THE WORLD.

Queen was one of the most memorable acts in the sixteen-hour, all-star Live Aid concert, a fund-raiser held in July 1985 in London and Philadelphia to raise money for famine victims in Ethiopia. Broadcast by satellite to billions of people around the world, Live Aid raised an estimated $125 million.[1] More important, it sparked a new trend of humanitarian reporting and celebrity fund-raising that brought world attention to the massive food crisis in Africa.

In the thirty years since, policy makers and aid groups have labored to improve rural livelihoods, agriculture production, and food security in Ethiopia. Yet the crisis is reemerging. By early 2016 over 20 million Ethiopians—one-fifth of the population—needed food aid, and that number was expected to swell by another million because of a strong El Niño–driven drought, a phenomenon that scientists say has intensified as a result of climate change.[2] El Niño is part of the El Niño–Southern Oscillation cycle, a weather system that can significantly influence rain patterns, ocean conditions, and marine fisheries across large portions of the globe. For much of the world the cyclical phenomenon brings drier-than-normal conditions—which is especially dangerous when a food system is already precarious.

Ethiopia has done much to improve its resilience to drought. In fact, the current drought is far bigger in that it is drier and more widespread than the one in 1984, yet the number of deaths from famine has not increased. The 1984 drought killed 600,000 people and gutted the country's economy. While Live Aid wasn't a long-term solution, it was enough to focus attention on the area until Ethiopia got back on its feet—when peacetime and rains returned. Writing in *The New York Times* in May 2016, Alex de Waal of the World Peace Foundation argues, "After countries have passed a certain threshold of prosperity and development, peace, political liberalization and greater government accountability are the best safeguards against famine. There is no record of people dying of famine in a democracy."[3] Today, even though a fifth of the population of Ethiopia is "desperately short of food supply," they aren't dying of starvation. Peace, transparency, and planning are all key factors in Ethiopia's resilience.

But global climate change, drought, and food shortages in the countries surrounding Ethiopia could have spillover effects that lead to setbacks. Aid agencies fear that international assistance to Ethiopians as well as the Sudanese and Somali immigrants in their midst could run out before the drought is over, primarily because aid is being diverted to other drought-stricken areas of the world, such as Syria.

Ethiopia makes a good case study because it carries the same bundle of food security risks, climate impacts, and other burdens as other sub-Saharan African countries. Its location puts it at risk for extreme weather such as drought and heat. Its social and natural systems are vulnerable to the stress of a growing population living on a finite amount of land. Ethiopia's population now ranks thirteenth in the world. By 2050 it is projected crack the top ten most populous countries. It has a wide range of ecosystems, and 75 percent of its residents live in rural areas where farmers depend on highly variable rainfall to grow crops. Rainfall is becoming more unpredictable with climate change, and El Niño exacerbated this effect. In 2016 two million people in Ethiopia were without drinking water for extended periods.[4] Like most droughts, this one was a slow-onset emergency.

The drought of 2016 followed multiple consecutive seasons of below-normal rainfall, making it the worst some parts of the country had seen in more than fifty years, and although El Niño was rapidly breaking up by the summer of 2016, Ethiopia was still suffering from its effects, which were expected to linger for some time. Food production fell. Livestock died. And the shortage of drinking water was acute. Ethiopia's projected need for food aid in 2016 was $267 million, according to the United Nations Office for the Coordination of Humanitarian Affairs.[5]

Images of hungry people, especially children, are unsettling. The LiveAid broadcast was full of pictures from Michael Burke's reporting for the BBC in Ethiopia in the 1980s—and they remain some of the most powerful depictions of food insecurity we've ever seen. But in time our memories faded. In fact, until the advent of the Arab Spring in 2011, the well-fed nations of the world seemed to have forgotten what it is like to be hungry. Then the Middle East refugee crisis brought it all rushing back. Suddenly, many of the global food relief agencies have become overwhelmed as they divert food and financial resources to Syria's drought-driven crisis and its refugees. Needy Ethopians have felt the decrease in support.

⊕

Researchers, development experts, and land-use planners have been fretting about hunger in sub-Saharan Africa for decades. Agriculture accounts for nearly half of Ethiopia's gross domestic product, according to the World Bank, including 84 percent of its exports and 80 percent of total employment. Coffee is among the leading exports, and it hit a record high of $862 million in 2015.[6]

The world is now in a transition from food abundance to food scarcity, writes Lester Brown in *Full Planet, Empty Plates*. Eroded soil, water shortages, and rising temperatures are some factors on one side of the equation. The 7 billion people in the world, projected to number 9.6 billion by 2050, along with rising affluence and the use of food crops as biofuel for transportation, are on the other. Brown argues that the move toward creating a resilient food system is only just beginning after decades of overproduction in the last part of the twentieth century.[7] Is it already too late?

Famine is the result of a triple failure: of food production, food access, and food policy. Despite strong efforts to improve access and production, climate change and population growth will exacerbate food insecurity even with strong disaster risk management and efforts to bolster resilience to natural disasters.[8] One path may be through improvements in seemingly unrelated areas: environment and health care. Population, health, and environment programs, as they are called, address the integrated but complex connections between human populations, improvements in health care, the planning and spacing of children, and environmental conservation.

Resilience is a new watchword for nearly every development agency as the threat of problems related to climate change looms larger. The population, health, and environment approach shares many of the principles of resilience planning, the broader set of principles used to help communities bounce back after shocks and stresses. Its integrated methodology, which reflects the complexity of real peoples' lives, offers new hope for improving food security in a changing climate.

The story of Tesema Merga, a village elder in Edibir, Ethiopia, brings the value of this approach into sharp focus. A long road outside his home, soon to be paved when he was interviewed, was expected to bring significant changes to his community. Today Tesema remembers how some of the first paved roads to Ethiopia's rural Gurage zone, a couple hours from Addis Ababa, transformed lives. "If you had to get someone to the hospital, it could take up to eight hours. Many patients died on the way," he explains in the opening segment of a film by the Wilson Center, *Paving the Way: Ethiopia's Youth on the Road to Sustainability*, which documents the efforts of the Gurage People's Self-help Development Organization. As detailed in the film, the organization, one of the oldest NGOs in Ethiopia, has adopted the integrated approach.[9] Eventually the roads connected the communities along them, and they not only improved access to health services but also allowed people to sell to larger markets, commute to jobs, and find new work.

But these roads cannot address some more recent challenges. "A combination of deforestation, localized population pressure, and rural poverty is preventing the region from continuing to grow in a healthy and sustainable way," the Wilson Center reports in *Paving the Way*.

Massive deforestation in Ethiopia has taken a toll on its soil. From a baseline of 40 percent forest cover in the sixteenth century, the country's wooded areas are down to 4.6 percent, a result of 0.8 percent deforestation each year, *The Guardian* reports.[10] Deforestation has led to substantial soil erosion, allowing more than a billion tons of soil to wash into the Nile every year. Rapid population growth contributes to this scarcity too. Regions like the Gurage rely on subsistence agriculture. Pressure on forests comes from a rapidly growing population, which ballooned from 48 million in 1990 to 101.5 million by 2016. The country's rural population has more than tripled, as rural women have on average 5.5 children during their lifetime. People younger than thirty now represent more than 70 percent of the population, according to the Population Reference Bureau. Over 80 percent of the population lives in rural areas and relies on rain-fed agriculture.[11] The 70 million livestock also put pressure on water, land, and forests.

Since 2008 the Gurage development organization has tried to address these interlinked challenges through its community development projects, such as erosion control, improving sanitation, and reducing deforestation, some of which were showcased in *Paving the Way*. Projects like these could help Ethiopia achieve the "demographic dividend"—a larger-than-usual ratio of working-age people to dependents that leads to increased economic productivity—by helping young people understand their reproductive rights and gain access to education and employment.

The Wilson Center reports that the need for family planning is high and unmet. "Nearly 75 percent of married women of reproductive age wish to delay childbirth, but only 28 percent are using a modern method to prevent pregnancy. And despite trends in the right direction, poverty remains widespread. A recent World Bank assessment found 37 million Ethiopians remain 'either poor or vulnerable to falling into poverty in the wake of a shock.' "[12]

The potential for using integrated approaches in Ethiopia is high, but in the absence of sustained programs and research to validate the value of offering reproductive options and environmental conservation programs, these approaches are difficult to replicate widely. The main peer-reviewed research on the validity of the integrated approach was published in 2010 and focused on the Philippines. Researchers documented multiple disciplines working together to produce results not obtainable by a single intervention alone, such as health services or environmental conservation. They found strong evidence that integrating population, health, and environmental programs outperforms single-sector approaches.[13]

When women in urban areas have access to family planning services and jobs, the fertility rate has demonstrably declined. Already, the population of Addis Ababa is at replacement levels, meaning that women are choosing to have two children, rather than the country average of 4.9. But options for women in rural areas—the vast majority—lag. People in rural areas have little access to health services, including family planning. In the South Omo Region of southern Ethiopia, where more than 700,000 pastoralists follow their livestock for grazing, the nonprofit Global Team for Local Initiatives (GTLI) works to improve the lives of indigenous peoples by using the integrated approach. Amarech Benki is a thirty-seven-year-old mother of nine children and a member of the Bena Tsemay tribe, one of sixteen tribes in the region. Her two oldest children are married, and the youngest is eighteen months old. Women in Benki's community usually have at least ten children.[14] She and her husband decided that, to make their family happy and healthy, she needed to use family planning, and she now receives a contraceptive injection once every three months.

The integrated program in her community first addressed its drinking water and sanitation needs, including pit latrines to prevent feces from contaminating drinking water. Next, the program focused on the nutritional needs of the families. Before participating in the program, the diet of Benki's family consisted of maize and occasionally meat when her husband's hunt was successful. As part of this Global Team program, she received four chickens when she graduated from the Integrated Functional Vocational Literacy school. She learned well, and she now has thirty chickens. She sold one chicken recently and used the money to purchase a book for her son. She has also given twenty fertilized eggs to five other households to help them start raising chickens. Her husband received vegetable seeds and agricultural tools when he graduated from the school. Their family's diet now consists of eggs, sweet potatoes, vegetables, chickens, and meat, which they buy at the marketplace with their egg money.

Benki's story is multiplied thousands of times over in Ethiopia, and other parts of Africa, as a result of the efforts of the population, health, and environmental programs. While many of the programs have failed to expand, because they're funded for only one to five years, others are finding ways to reach more people. The PHE Ethiopia Consortium is pooling resources and funding to help coordinate and plan population, health, and environmental programs throughout rural areas of the country. The mission of the nonprofit consortium is to improve rural residents' understanding of the connections between a growing population, the state of the environment, and the availability of work. The consortium consists of dozens of local and international NGOs that are working together to help the government meet the goals of Ethiopia's latest five-year economic and growth transformation plan, which received $50 million from the Green Climate Fund, a multinational organization that is accountable to the United Nations, for climate-resilience projects based on sustainable use of natural resources.[15]

A biodiversity hot spot known for its critical wetlands and more than 200 species of fish, Uganda's Lake Victoria Basin is experiencing environmental degradation from climate change, agriculture, pollution, overfishing, and increasing industrialization. The basin is home to 42 million people and stretches from Uganda and Kenya to Tanzania, Rwanda, and Burundi. More than 80 percent of the basin's population relies on fishing and agriculture for sustenance. The integrated programs in the area address the interlinked threats to ecosystem conservation, capacity to manage natural resources, and sexual and reproductive health.

One of the more provocative problems in the region is Lake Victoria's gendered fishing economy and the "fish for sex" phenomenon. Men fish. Women are expected to purchase the fish from men and then dry and sell the catch at market. With small-scale fisheries in decline, however, fewer fish make it to shore. "As demand increasingly dwarfs supply, female vendors report a need to garner favor with these fisherman beyond payment of the selling price itself in order to obtain a sufficient supply of fish to sustain their livelihood. As a result, informal relationships develop and are sustained by transactional sex as supplemental currency for fish."[16] (Among other problems, 22 percent of the basin's population is HIV positive.) Climate change is likely to further reduce fishing stocks in Lake Victoria. The so-called hot fish research by Michael Cooperman and others indicates that inland fisheries are less likely to be able to adjust to temperatures that are rising as a result of the changing climate. (For more on hot fish, see chapter 9.)

Poverty and poor access to health care collide with food security and conservation beyond East Africa. Consider Madagascar, another biodiversity hot spot. An estimated

80 percent of its flora and fauna are found nowhere else in the world. But its population is growing rapidly, expected to double by 2040. Communities there rely heavily on local natural resources, especially fish, to make their living. The country has extremely high levels of poverty. Ninety-two percent of Madagascar's population lives on less than two dollars per day. Food insecurity and poor access to basic health services strain households even further.[17]

While some population, health, and environmental programs start with a focus on integrating population and family planning to achieve a conservation goal, others do not. Blue Ventures, a London-based NGO, came into that work in reverse. While the organization has been working on integrated programs in Madagascar since 2007, its initial goal was to provide viable alternative livelihoods to fishing through seaweed and sea cucumber farming, and to educate communities about sustainable management of natural resources. Blue Ventures started on Madagascar's southwest coast. "Madagascar has a very wide range of habitats, both terrestrial and marine, and these are all experiencing significant degradation," said Caroline Savitzky, a community health program coordinator for Blue Ventures. But the organization soon found a tremendous need for contraceptives. In response to an overwhelming demand for health services, Blue Ventures added reproductive health services to its natural resource management and livelihood programs. "In some communities people had to walk the length of a marathon [twenty-six miles] to reach basic care," Savitzky said. "About three-quarters of a million women in Madagascar want to be able to plan their families but are not currently using modern methods of contraception."[18]

Blue Ventures went outside its comfort zone to address the need for family planning, but doing so was clearly essential to achieving the larger goals of natural resources conservation and biodiversity. Conservation organizations are often hesitant to work in health care, for fear of "mission drift," but Savitzky said a synergy exists between the two. Population, health, and environment programs go beyond food security because they target ecological hot spots, areas that are richest in biodiversity. "While these hotspots [*sic*] comprise just 12 percent of the planet's land surface, they hold nearly 20 percent of its human population, with little access to basic government services like health and education," according to an environmental health website supported by USAID. This presents a big opportunity for the programs to both build in resilience and improve health and food security, especially because that hot spot–based population is growing nearly 40 percent faster than that of the world as a whole.[19]

After establishing itself in both the conservation and public health realms in Madagascar, Blue Ventures eventually formed a partnership with the global sexual and reproductive health organization Marie Stopes International, and the Duke [University] Lemur Center, to better integrate health and animal conservation efforts in north-eastern Madagascar. The partnership also helped found the Madagascar PHE Network, which brought together thirty-five health and environmental organizations.

While integrated programs can provide an element of resilience to communities facing sweeping change, other dynamics are at play. Notably, when populations grow so much that competition for land increases, food security can spur conflict. Population and food production have been on Robert Engelman's mind for decades. Engelman, author of *More* and senior researcher at Worldwatch, wrote about that problem back in 1980, when he was an environment reporter for the *Kansas City Times* and already making the connections about the growing U.S. population and the limits of agricultural land. "The thing about land is that they ain't making any more of the stuff," Engelman said, quoting Will Rogers.[20] In 1980 the world population was 4.4 billion, a mere three-fifths of today's 7.3 billion.

Engelman argues that, as the population soars, the way we feed and care for ourselves will become unpredictable. He challenges the presumed one-two punch of economic growth and population growth amid climate change, saying the possibilities have not been fully explored. (With population growth steadily rising, we will need to produce more calories to meet the nutrition requirements of more people. With economic growth, we will need to produce more meat as more people—especially in China—adopt a middle-class "Western" appetite. Population projections are demographers' current "best guess," he said. They could be much higher with no end in sight, or lower, if family planning solutions are quickly adopted.

The other big variable is the consumption of natural resources. Consider Lake Urmia in northwestern Iran, which is disappearing. Already, about 80 percent is gone, totally destroying the ecology and the tourism industry of the region. Researchers have examined the history of precipitation, drought, temperature, and uses of water around the lake and have concluded that climate change was not the culprit, as many people had suspected. Rather, the culprit—overwhelmingly—was inefficient and misguided use of the streams feeding the lake and the water of the lake itself.

"The same thing plays out in lots of places," Engelman said.[21] As I mentioned in chapter 4, the tens of thousands of refugees streaming into Europe in the wake of Syria's civil war are largely fleeing intense violence and a crisis that owed its early beginnings to drought. What isn't often reported is that the population of Syria more than quintupled, from four million in the 1950s to 22 million just before the civil war broke out. Through mismanagement and overuse Syria had largely exhausted its groundwater reserves, so when the droughts hit, the water supply was grossly insufficient to withstand them.

Around the world, and especially in Africa, fertility rates are dropping—not as quickly as experts wish, but they are coming down. Fertility rates are closely related to the power that women have over their own lives and the commitment that governments make to encourage women to manage their own lives and to make their own decisions. One important element is the ability of women to make choices about if, when, whether, and with whom they have a child. When explored in the context of food security and natural resource conservation with the question "Are we exhausting the planet?," access to family planning becomes an important part of the integrated approach to improving food security.

India's Climate-Smart Villages

The village of Dhundi lies far from the paved road and even farther from anywhere else in India. The road stretches between the towns of Dakor and Ambav, an hour northeast of Anand, in central Gujarat. Few visitors venture into this agricultural region. Like most places in India, navigation is an adventure. Drivers pass camels pulling wooden carts filled with coconuts, and teams of oxen draw wagons filled with forage for dairy cattle. Speeding trucks, buses, and three-wheeled motorized rickshaws honk liberally, as cattle and goats clog the margins. And if you leave the main road, all signs of modernity quickly fall away.

At the first bend in the rutted route to Dhundi, a male

elder wearing a white cotton loincloth was sitting cross-legged on the gravel at the edge of the road. He was tending to his goats and buffalo, which grazed on nearby shrubs and grass. My driver eased into the turn to give the man a wide berth. A few hundred yards farther on, three gray-haired women dressed in colorful saris were reclining on traditional charpoy rope beds in open-sided huts. The road itself was crumbling and riddled with potholes; I immediately wondered what it must be like during monsoon season. It floods often, one villager confirmed.

Dhundi village is perhaps the quietest corner of Gujarat State. Some tourists visit the region for its wildlife sanctuaries and holy places, but the treasures of Gujarat are diverse and mostly hidden. Agriculture covers 60 percent of the land area, and large-scale industry—processing cotton, sugar cane, and dairy products—makes it a major industrial hub as well.[1]

I went to Gujarat to learn what makes a climate-smart village—the reputation Dhundi has earned in the environmental community. Energy reporting from this part of India has been dominated by stories about large-scale solar investments, canals to move water, and depletion of groundwater. The national government and international water advocacy groups have made repeated efforts to limit power to irrigation pumps to only a few hours per day to encourage farmers to use less groundwater and fertilizer in some areas and increase irrigation in other areas. But these efforts to encourage conservation have largely been failing. I had come to look at the latest innovation at the energy-water nexus, a solar-powered water pump that was about to solve a big problem for the people living at the end of this long, dead-end road.

On the day I visited Pravin Bhai Parmar, a twenty-nine-year-old who lives in Dhundi in a modest two-story brick home with his wife, two children, and extended family of eleven, he asked me to write my name in a notebook that serves as his log of guests. I was the first entry. He bought the notebook to keep track of the visitors he expects will arrive in the coming months because they will want to see the solar panels in action, he said.

I sought Parmar out after learning that he and his neighbors had recently joined forces to install solar panels in their fields to power their irrigation pumps. Two months earlier they had made a down payment of 5,000 rupees (US$80) each. It was the first step in making Dhundi's farming practices climate resilient. The farmers will ultimately invest another 45,000 rupees (US$700) each—a small fortune to them—to get the project up and running.

I sat with Parmar and his coinvestors—seven other young farmers—in his open-air living room. While the houses in the village have electricity, service does not extend to the field around back, he told me. The 1,500 residents, including Parmar, are mostly classified as Other Backward Caste, a socially and educationally disadvantaged group. The Parmar family's brick house has three rooms on the first floor, but it is unusually big for the village. The interior room, or living room, has three walls and a traditional layout. It opens onto the home's small courtyard, where the family shelters its livestock under a partial overhang. As we drank thick masala chai tea from small glass cups, we looked out at the stalls, home to two water buffalo that were nibbling on freshly cut greens while chained to hooks on the wall. One cow urinated on the cement floor.

Parmar explained the challenges of growing his own food. On less than two acres, his family grows vegetables, rice, millet, and wheat, and they rotate their production seasonally. The crops in this region rely on rain during the summer monsoon, which begins in June and ends in September, roughly speaking. The rest of the year they rely on irrigation. Parmar hoped that with the help of the solar panels, he will be able grow more food all year long, especially vegetables, such as eggplant, that bring a higher value at market. When we spoke, he could not grow vegetables outside the summer months because the cost of irrigation was too high, he said.[2] Instead, he was growing millet, which tolerates drought, to feed his family.

But millet has its problems. Despite its high nutritional quality and strong climate resilience, it carries the stigma of food for the poor. It isn't that tasty. And while it does not need much water to grow, harvesting it is much more labor intensive than wheat and rice. Millet is difficult to process and grind into usable flour, and it takes longer to cook, which means it consumes more fuel, Parmar said. And if he has any surplus millet to sell, it brings a much lower price at market than vegetables would.

Irrigating crops without electricity has always posed a

serious challenge for Parmar. To run his diesel-powered water pump, he and his brother had to drive on a motorcycle to the nearest town, about a twenty-minute trip each way, fill jerrycans with fuel, and then drive the heavy load back to the village. The process was both time consuming and expensive. The diesel cost amounted to 750 rupees per day, or US$11, which ate significantly into his crop profits. Diesel for irrigation can cost 15,000 to 20,000 rupees per year (US$220 to $295).

Rainfall in India has been erratic since about 2005. When it finally arrives, it comes in a few heavy torrents. Heat is stressing the crops. Groundwater used for irrigation is increasingly depleted. So how can solar help? In addition to energy, it provides a financial incentive for farmers to conserve water because they can sell excess energy back to the grid, thus easing the strain on depleted aquifers.[3]

"When you connect the solar pump to the grid and let the farmers use the energy they need for the pumping, and you give them the chance to sell the surplus solar power to the grid at an attractive price, then they will opt to do it," Tushaar Shah, senior fellow at the International Water Management Institute, told me. Solar pumps are part of an entire system of adaptive measures being implemented at roughly eighty test sites, called climate-smart villages, across six Indian states. The organization responsible is the international agriculture research group CGIAR (it uses only its acronym), whose research program on Climate Change, Agriculture and Food Security is working with Shah's institute and other NGOs to provide financial incentives for farmers to be more climate-clever.

I met with Shah one morning in the water institute's field office in Anand, Gujarat. He denounced the current government subsidies, which give farmers little incentive to limit their use of electricity—or, worse, diesel-fueled irrigation pumps. "The solar energy will give the farmers a crop [solar] that is worth up to 90,000 rupees a year. We think this will reverse the current incentive structure that has led to overpumping. There are very few crops [that] farmers will grow that give you that kind of income," he said.[4] Luckily, climate-smart technologies like the solar pump are slowly gaining acceptance among village communities.

About a half hour by car from Dhundi, along another lumpy dirt road that skirts the village of Thamna, solar panels were driving a water pump that irrigates the fields of farmer Raman Bhai Parmar, sixty-five (no immediate relation to Pravin Bhai Parmar). Raman Parmar grows bananas, rice, and wheat on seven acres. We sat under a mango tree and talked about the challenges of farming when the monsoons are unpredictable. The whooshing sound behind me came from a water pump connected to a concrete holding tank that he used to irrigate banana and rice crops in the adjacent fields.

As we looked out over the rice paddy, a wooden oxcart plodded by on a rutted dirt road; the cart was loaded with handpicked cattle fodder tied in bundles with scraps of fabric. I was struck again by how unlikely it is that a place like India is leading innovations in agricultural resilience and efforts to reduce greenhouse gas emissions, although the country may have more farmers per capita than most places in the world. But India may not have much choice. Growing food in the region is already increasingly punishing. Farther north of Gujarat, in climate change hot spots like the states of Haryana and Punjab, studies are showing that yields of wheat—which is particularly vulnerable to heat stress—will decrease 6 percent to 23 percent during the next three decades.[5] Such unpredictability is persuading farmers to reenvision their livelihoods. Although society needs to produce more food to meet future demand, agriculture is a major driving force of greenhouse gas emissions, water-quality degradation from groundwater use, soil loss, and nutrient runoff on the planet.

Farmers—both in India, where over 70 percent of the population still is economically dependent on agriculture, and the world over—can no longer idly depend, as they have been, on small scientific breakthroughs from time to time. In India neither breeding higher-yielding varieties of its dietary staples (wheat and rice) nor perfecting irrigation methods will come close to sustaining the country's growing population, projected to increase by another 346 million people—more than the current U.S. population—by 2050. Farmers instead must now be intimately in tune with their local ecosystems, whether the climate-smart model or other interventions that consider the local interconnectedness of weather, water, nitrogen, carbon, and energy. The climate-smart concept is an approach for transforming and reorienting agricultural development under the new realities of climate change through interlinked goals of sustainably increasing yields, enhancing resilience, and reducing or removing greenhouse gas emissions where possible. As farmers and researchers are beginning to understand, fixing one problem—say, food security—in isolation from the others—agriculture and climate change—cannot solve the overall problem because even seemingly minor details can easily undermine the entire system.

A cohort of researchers from the water institute and the International Wheat and Maize Improvement Center, working with CGIAR's Climate Change, Agriculture and Food Security program, is trying different concepts in each of these climate-smart villages, which currently number 1,500: 500 in Haryana, 500 in Punjab, and as many as 500 others throughout the country, including Bihar and Karnataka.[6] Much is riding on their success.

To see examples of climate-smart practices in other regions, I traveled by car for a day to India's northern border with Pakistan to visit the Punjab village of Noopur Bet outside Ludhiana. There I met Joginder Singh, sixty-eight, who was starting to adopt new sustainable farming practices in his fields. Singh remembered what it was like to grow up in the aftermath of a famine.[7] In 1943 several million people died in Bengal, India. It was a jolt even for a country whose history is marked by brutal and repeated famines, droughts, and food shortages.

After that tragedy—and after India achieved independence from Britain in 1947—the government began to focus especially on the challenge of feeding its people. Building on the work of Norman Borlaug, the plant breeder from Iowa known as the father of the green revolution, Indian scientists improved their breeding of wheat and rice; irrigation; and the use of chemical fertilizers. As a result India no longer relies on massive food aid, which had consisted mostly of wheat and cereal imported from the United States.[8] But as the country's population further swells, its small farmers will be hard pressed to maintain yields.

Andrew Jarvis, a senior scientist and program leader at the International Center for Tropical Agriculture in Cali, Colombia, is working with the Climate Change, Agriculture and Food Security program to coordinate research on the climate-smart concept globally. "We see adaptation as joined-up thinking. It doesn't work when you have a maize specialist doing his own thing," Jarvis said. "This is about working together with farmers and markets within local conditions and institutional capacities sharing knowledge."[9]

In 1960 food aid was 92 percent of the annual U.S. foreign assistance budget for India. From 1992 to 2002 food aid accounted for only 65 percent. The food aid program ended in 2011, largely because of gains made in the small northern breadbasket states of Punjab and Haryana. Those two regions make up a tiny fraction of India's land area yet supply almost a fifth of the country's rice and more than a third of its wheat.[10]

On my visit to India in September 2015, a few weeks

before the rice harvest, the vast green expanse of flat fields and farming villages looked much like the southern Minnesota farmland where I grew up. In late summer Minnesota bulges with soybeans and corn, while India swells with rice in the fall and wheat in the spring.

The fields were a reminder that strategies once critical for boosting food security in India were first honed in the American Midwest. Between the 1950s and the 1980s U.S. crop yields improved by 3 percent year after year, an accomplishment that's no longer realistic.[11] As India developed its agricultural capacity, it experienced tremendous gains by growing a high-yielding variety of dwarf wheat introduced by Borlaug.

While in Delhi I went to see a bronze memorial to Borlaug at the Indian Agricultural Research Institute. Around the statue's neck was a garland of marigolds, which represent passion and creativity in India. A biologist by training, Borlaug received the Nobel Peace Prize in 1970 for his innovations in high-yielding wheat. In both the United States and India, growing high-yield crops using Borlaug's methods improved significantly how much food farmers could grow and harvest from their fields. After farmers adopted Borlaug's wheat, their standard procedure was to apply large amounts of fertilizers, water, and pesticides to ensure plants grew and thrived. But the increases in harvests eventually leveled off and extracted an ecological cost.

In India the production methods of the green revolution are no match for climatic variability. Studies project that yields of wheat will drop 6 to 23 percent by 2050.[12] As a result the Borlaug Institute for South Asia in Punjab, which works with the climate-smart villages, is now researching new approaches. I visited the institute and saw test plots where researchers are interplanting mung beans and wheat to sequester carbon and improve organic matter in soils. They are testing water efficiency through underground drip irrigation for corn and rice. They are also using conservation agriculture techniques to reduce the methane produced by growing rice.

In the past the world looked to U.S. grain stores as a backup supply of food during extreme weather, disasters, and shocks. But as shocks and disasters have grown more frequent, the stocks have dwindled.[13] Grain stores may have trouble keeping pace with India's projected food shortages. Neither breeding new varieties of wheat and rice to achieve higher yields nor perfecting irrigation methods is likely meet the wheat and rice needs of India's hungry and growing population, which is also shifting its patterns of consumption.

The near-term outlook in the United States is somewhat mixed. Between 2016 and 2040 many American farmers are expected to be fairly resilient in the face of climate change even as other parts of the world experience big crop losses from stress as a result of extreme heat, drought, disease, and heavy downpours. Certainly, climate disruptions to agriculture have increased in the United States since 1975 and are projected to continue through 2040. Those impacts have been regional, and regions will continue to experience declines in crop and livestock production from increased stress due to weeds, diseases, insect pests, and other climate change–induced stresses, according to the U.S. National Climate Assessment. But other parts of the world will likely fare worse.[14]

Punjab's Joginder Singh told me he has now started to use new methods of farming to improve—or at least maintain—crop yields despite climate change. With his son, Singh has used a laser-guided tractor, which he leased from the local farmers' cooperative, to flatten his fields with precision. Flatter fields improve water efficiency, he said. When he adopted climate-smart interventions like laser-guided leveling of his fields, he conserved 20 percent of the water resources his fields had formerly used and increased his yields by 15 percent through greater precision in seeding, tillage, and measuring the moisture of his soils before irrigating. Some farmers use a handheld crop sensor called a Green Seeker to assess crop health, and mobile phone apps calculate how much fertilizer to apply throughout the growing season. Singh also avoids tilling his fields, which helps the soil retain moisture and leads to fewer costs and fewer greenhouse gas emissions.[15]

Another problem with farming in India is the overapplication of fertilizer, which is inexpensive because of government subsidies. But adding too much has led to increased greenhouse gas emissions, soil contamination, and polluted groundwater. To address that, Singh's son uses a digital application on his smart phone and a simple color chart to calculate and time how much and when to apply fertilizer and irrigate fields, improving water and fertilizing efficiency as well as strengthening yields. The father and son also listen to voice messages of detailed weather forecasts delivered on their mobile phones, before they plant and throughout the season, to determine when the monsoons will occur—the timeframe for which is becoming increasingly variable because of climate change.

In July 2015, when a third of the total rainfall the region typically receives in an entire season fell in only a few days, many fields became waterlogged because the soils could not absorb the inundation. Many crops in the fields of some of Singh's neighbors suffocated and died. But because he was practicing climate-smart interventions, and did not burn or till the residue from the last wheat harvest (something farmers typically do to prepare for the next crop), the flood did no harm; he had seeded rice directly into the field the month before. And because his field held more organic matter, the soils absorbed the water and the crop survived. "I was worried for a month until I saw the rice sprout," he said. "It wasn't until after the heavy rains that I knew the system would work."[16]

Punjab and Haryana, the grain basket states of India, produce the majority of the country's staple wheat and basmati rice for export to the Middle East and European markets. Since the mid-1980s the pumping of groundwater for irrigation to produce these crops has led to a spike in productivity and increased food security. However, the region faces future threats from heat as well as to its water resources. Wheat is particularly vulnerable to heat stress, and temperatures are predicted to increase by as much as 9 degrees Fahrenheit, or 5 degrees Celsius, by 2080.

When I met with Vikas Chaudary, thirty-four, in Tararori, Haryana, he told me that he had learned farming from his father. When he first got into the family business, rains came predictably during the monsoon season and agriculture was a safe bet for a career. Groundwater was plentiful. Soils were rich. Now all that's a gamble for Chaudary, who farms thirty-five acres and grows rice and a small plot of corn in summer and wheat in winter to support his extended family of seven.

I met Chaudary at a demonstration field near the city of Karnal. He was giving a talk to international business leaders from such companies as the global agribusiness Olam and the accounting firm PriceWaterhouseCoopers, who wanted to understand sustainable agriculture. That day the field temperature was so high that one of the accountants fainted.

Climate-smart farming isn't easy. The cost of the specialized machinery needed to plant the next crop in the rotation amid the stubble from the previous season is a barrier for small farmers, Chaudary said. Many villages have farming cooperatives that share the cost of equipment, but the cost is still out of reach for the majority of farmers, who are accustomed to the conventional technique of planting in bare soil. I was most surprised to learn that many farmers have had a hard time convincing their wives of the need to change their farming methods—by adopting no-till farming, for example. While not tilling fields or burning residue after harvest has distinct climate benefits and organic matter improves soil, the field can appear messy compared to a neatly tilled plot. Some men told me their wives thought they were being lazy by not tilling the fields. Some wives even withheld meals from their husbands until they agreed to run the plow. The farmers eventually asked the field technicians who specialize in climate-smart agriculture to speak with the women to explain the benefits and reasons for no-till methods. This discussion prompted the researchers to actively include women in the planning of climate-smart villages.[17] The man leading the research on climate-smart practices in India is Pramod Aggarwal, regional program leader at the water institute in Delhi. With colleagues at the International Wheat and Maize Improvement Center office, also in Delhi, he's helping to introduce climate-smart villages to South Asia and beyond by providing models to countries whose food production is acutely vulnerable to climate change. Despite the looming threat, all the researchers with whom I spoke were optimistic—especially Aggarwal, who seemed to be a scholar at heart with a quick wit and a kind smile. Briefing world agribusiness leaders on climate-smart developments one minute and joking with farmers the next seemed to come easily to him. Success, he told both groups, hinges on landscape-level interventions—essentially, re-engineering how entire systems work together. I watched as Aggarwal and colleagues, along with Chaudary, led business leaders and farmers on tours to demonstrate the advantages of this approach. Aggarwal is well aware that farmers are, first and foremost, businessmen, so he appeals to their economic sensibilities, which works. The farmers are seeking to adopt these new methods not just for the environment's sake, he told me, but because it makes fiscal sense.

One might assume that the same package of interventions could help climate-smart villages increase proportionally throughout the world, but in fact the opposite is true. Such interventions may take many other forms, depending on local livestock and cropping systems, natural resource constraints, markets, and social and gender dynamics. "The emphasis is always to dovetail with local conditions, local institutions, local governments. These are not discrete changes," Aggarwal said. "Our key goal is to raise the capacity in agriculture—whatever developments are done should not lead to any maladaptation in the future. Crop yields should be able to be sustained and should not contribute to higher emissions."[18]

The future of Aggarwal's work in South Asia—and beyond—is promising. Already, leaders in the parched eastern state of Bihar and the monsoon-dependent state of Karnataka were committing to more climate-smart villages. In early September 2015 a group of fifty researchers, leaders, and policy makers from around the world gathered in Punjab to plan how climate-smart villages in other developing countries can help meet the needs of farmers, including women and marginalized groups.

Farther afield, eighteen developing countries around the world are tailoring the concept to fit their locales, including Burkina Faso, Ghana, and Senegal in West Africa; Ethiopia, Kenya, and Uganda in East Africa; Colombia, Honduras, and Nicaragua in Latin America; and Vietnam and Cambodia in Southeast Asia.[19] The ultimate value of climate-smart villages, Aggarwal said, lies in raising "the literacy and capacity of farmers to address the complexity of climate change" and showing them "the solutions that are possible if they organize themselves as communities."[20]

◍

Climate-smart villages are, at heart, a systems engineering concept. The variety of systems under the banner varies greatly and cuts across all areas of farming.

Rice is a focal point for climate-smart practices around the world for a few reasons. First, it's the primary source of nutrition for billions of people worldwide. And it's a thirsty crop. But growing it in flooded paddies—a common technique—means it is a main source of methane, a powerful greenhouse gas.

Alberto Mejia spent three years tinkering with climate-smart growing practices. Water is a main concern in regard to changes in climate, and he's trying to reduce the amount of water he uses for irrigating his 1,100-acre farm near Ibague in the tropical central range of the Colombian Andes. He now plants varieties of rice that require less water. He floods his paddies with greater precision and has installed gauges that measure the moisture content of the soil. On a daily basis he can determine how much nitrogen the plants need, and he relies on more advanced weather forecasting to plan when to fertilize, water, and harvest the grain.

"We are learning how to manage the crops in terms of water, which will be a very, very good help for us now and in the future," Mejia said, adding that the El Niño weather pattern had caused serious drought in 2015. "We have very difficult days—hot, with no rain. It's dry. There are fires in the mountains. Growing crops makes it a complicated time here."

Ever since a drought devastated his yields in 2009, Mejia has been eager to make sweeping changes to his rice production. He believes that the weather has become more erratic and is concerned that future climate change will make rice farming even more difficult. As a result, and with the help of his local rice growers' association and scientists from the International Center for Tropical Agriculture, he is embracing climate-smart agriculture.[21] These are agricultural techniques that protect farmers from the effects of global warming and improve crop yields while also limiting greenhouse gas emissions.

The move to climate-smart agriculture is strongly supported by dozens of organizations such as the World Bank, the United Nations Food and Agriculture Organization, and the CGIAR Consortium, a network of fifteen international research centers that work to advance agriculture research globally. The Global Alliance for Climate-Smart Agriculture, an organization of governments, international and regional organizations, farmer's organizations, and civil society groups that began work in 2014 to establish

The author at the motorized picnic table in Vermont at the start of the journey to understand the many paths to a sustainable food future.

The author's colleagues from Vermont's Environmental Law Center at the motorized picnic table in the alfalfa field.

Vermont's communities, landscapes, and policies favor local food.

Cattle nearly hidden amid thick fields of silvopastoral grazing at El Hatico ranch in Colombia.

Carlos Hernando Molina, owner of the Colombian ranch El Hatico, explains how shade and the dense brush of silvopastures help retain soil moisture and nutrients and lead to evenly distributed deposits of manure.

Fabiola Vega started raising livestock with agroforestry methods in 2012 after she heard a presentation at a community meeting about how to sustainably raise cattle. Her farm, La Cabaña, is located in the Andean foothills of the northern Valle del Cauca in Colombia.

Many crop irrigation systems, spray irrigation in particular, are inefficient and have been called evaporation machines. This rig in Oklahoma draws from the quickly depleting Ogallala Aquifer. Improving efficiency in water irrigation will become a priority because of climate change.

In the Kasese district, Uganda, households build a greener—and more food secure—future through sustainable livelihoods. Researchers studying climate change, agriculture, and food security conduct individual interviews about intra-household decision making with smallholder coffee-farming households in order to better understand gender equity and differences in bargaining power. (Photograph by Els Lecoutere, CIAT/CGIAR)

Resilience to a hot planet starts with a good harvest in villages in Nyando in Western Kenya. Since 2011, researchers, development partners, and government extension agents have been working in Nyando to develop a mix of technologies tailored to boost farmers' ability to adapt to climate change, manage risks, and build resilience. These technologies are expected to in turn improve livelihoods and incomes. (Photograph by S. Kilungu, CIAT/CGIAR)

The author with the Norman Borlaug statue at the National Agriculture Science Complex in New Delhi, India. The statue is regularly strung with garlands of marigolds as a sign of respect.

Joginder Singh (right), a farmer in the Indian village of Noorpur Bet in Punjab, and his son are among the thousands of farmers in India trying to reconcile the risks posed by a changing climate and their need to improve crop yields to support their families.

Joginder Singh stands in his rice paddy near the Indian village of Noorpur Bet, Punjab. He started practicing climate-smart agriculture this year. He does not read or write, but he learned the methods after visiting a demonstration site.

The author (left), speaking with Raman Bhai Parmar at his farm in India. Water, energy, and food security are inextricably linked. Here, in a village one hour north of Anand in Gujarat, Parmar uses solar panels to pump water to irrigate his banana, rice, and wheat crops.

At the Borlaug Institute for South Asia in Punjab, India, agriculture experts are searching for new ways to grow crops in what they expect will be a water-stressed area in the future. Here, Dr. Harminder Singh Sidhu tests plots of corn grown with drip irrigation.

This vast field of corn (left) in Paraguay grows amid what was once a dense section of Atlantic Rainforest. Now only a few trees survive. (Photograph by Holger Kray, World Bank)

Verlasso salmon is a farmed fish that is marketed by touting the company's sustainability practices. Verlasso has reduced its reliance on wild-caught fish for salmon food.

Scott Sibbel grows crops and livestock on his farm in west central Iowa. He earns a premium for his conservation farming practices and humane treatment of livestock.

support for climate-smart practices, wants to strengthen global food security, improve resilience to climate change, and help 500 million small farmers adapt to more stressful growing conditions.

Crucially, climate-smart agriculture helps farmers and economies adjust to the stagnating yield gains experienced in the green revolution, particularly with rice and wheat. Using seeds specifically bred to withstand certain temperatures or moisture levels, coupled with better water management, can help to keep improving agricultural productivity, if not at quite the same clip. For example, in Rwanda projects include better management of rainfall on steep hillsides and terracing that prevents water runoff and erosion. In Senegal various organizations are providing planting, growing, and harvesting information to women, who do the majority of farming but historically have not benefited from agriculture extension services because communications have focused on the cash crops that men tend to grow and bring to market, such as corn, sorghum, and millet. The women receive text-message alerts and information on blackboards at community outposts to provide them with advice on seeds, fertilizer, planting methods, or weather patterns that affect the kitchen crops women commonly cultivate, including rice, tomatoes, and onions.[22]

Colombia's farmers learned a hard lesson in 2010 and 2011 when drought, high temperatures, and acute water shortages devastated crops. That's one reason the country's agriculture ministry, farming organizations, development agencies, and researchers sought ways to improve resiliency to climate change, especially because it is predicted to increase weather variability. Researchers at the International Center for Tropical Agriculture teamed up with Fedearroz—the Colombian rice growers' association—and the Colombian Institute of Hydrology, Meteorology, and Environmental Studies to analyze climate and rice production patterns in selected regions of Colombia. Jarvis, of the international agriculture center in Colombia, said that through analysis of big data, researchers and trade groups can provide rice growers with specific recommendations to improve their production practices and avoid the worst effects of climate variability.

For farmers like Mejia, who plant new fields each month to maintain a continuous cash flow, shifting production to a certain period each year runs counter to what farmers learned from their fathers and grandfathers. But increasingly unstable weather in recent years has made many Colombian farmers more willing to try new ideas. "They realize that climate change is a long-term threat," Jarvis said. "In the short term, it is showing itself as climate variability, and so we need to adjust to it."[23]

Mejia is a numbers guy and keeps careful records. Most of his recent agriculture modifications, such as planting new kinds of rice and flooding the paddies only when the humidity meters say they need water, have proved successful. Each day he checks how much nitrogen the plants need and consults the weather equipment at the farm to learn how much rain has fallen, the wind speed, and the high and low temperatures. Most important, he reviews a forecast that now extends to eight days.

Previously, his rice paddies took three to four days of flooding before he drained them. But the rice growers' association told him he could get away with flooding the fields every five to eight days instead, if he measures the moisture levels in the soil. Still, his yields have dropped 30 to 40 percent in the last two years because of drought. As a result he has started planting fewer fields during droughts, since his investment in seeds and fertilizer will almost surely be lost, he told me.[24]

Another Colombian farmer, Oscar Perez, plants 320 acres of rice annually, in addition to corn and cotton, on his farm in the hot and humid area of Cordoba near the northern Colombian coast on the Caribbean. Following the advice of Fedearroz and the international agriculture center, Perez has begun not planting fields to avoid losing seed and incurring the expense of fertilizers and labor costs that cannot be recovered.

"When there is no rain, you just don't produce," said Perez, who has operated his farm for fifteen years. He has begun taking advantage of better weather forecasting and avoids applying fertilizer when heavy rains are predicted so it will not be washed away. In addition, he is using better seed selection to gain improved yields for conditions that are determined to be wet, dry, or moderate that season.[25]

In Vietnam farmers now rely more on chemicals to

control weeds, so they don't need to flood their paddies as often to use water as weed control. The use of chemicals can contaminate the water, but the fertilizer spares the rice crop from the debilitating effects of drought by intensifying production and increasing yields. As a result farmers in Vietnam have reduced the amount of seed used per field by 70 percent, reduced water use by 33 percent, and cut the application of nitrogen fertilizer by 25 percent, according to CGIAR research. In addition to closely monitoring water levels in the soil, the farmers use high-yield seeds bred to withstand the occasional soaking of saltwater by rising seas.[26]

Experts suggest that climate-smart villages provide an important means for building resilience to climate change, because the village approach breaks through traditional barriers and addresses both social systems and agronomy. It's already clear that new methods of agriculture may not suffice in a world of rising temperatures and soaring populations, especially in places like Africa. Simple interventions at the farm level can go only so far if water is unavailable or soils degrade beyond repair. It's necessary, experts say, to look beyond the farm and manage entire landscapes that support people, food production, and nature.

Seth Shames, director of policy at the nonprofit group EcoAgriculture, said that preserving the ecosystems of an area—including forests and wetlands—is one essential part of supporting agriculture. But he warned that if serious efforts are not made to slow climate change, all the work be for naught.

"If projections are correct, the destruction that will occur in agriculture will be so severe that those kinds of solutions will be swamped by reality and God knows what will happen," Shames said. "In twenty to twenty-five years we will get to a point in some places that either it will be too hot, too dry, too wet, or too cold for the crops you are planting, and you will have to put something else in its place, which will be incredibly disruptive at best."[27]

Sustainable land use in agricultural systems under climate change means that agricultural practices will need to be implemented at the ecosystem level. This means farmers and governments will need to coordinate their decisions across political boundaries and much larger areas, such as valley to valley (watershed to watershed) or mountain range to sea, than is typical. These collaborative decisions will have to take into account much broader goals for ecosystems' health relative to agricultural output. This practice stands in contrast to standard decision making, but increasing numbers of farmers are likely to take it up as climate change continues.

Climate Stupid

Holger Kray was standing at the edge of a field of soybeans in Paraguay and eyeing the last remaining tree in this section of the Atlantic Forest. Until recently the field was part of a tropical rainforest that stretched along the entire coast of Brazil and inland to Paraguay and Argentina. Now, over 85 percent of it was gone.

Kray is a lead agricultural economist at the World Bank, which means he analyzes the mathematical side of sustainable farm practices in rural development. He took a photo of that tree while visiting Paraguay to try to nudge farmers toward converting to a sustainable agricultural economy. As I looked at the black-and-white image, I noticed the stark antiseptic landscape surrounding the tree that is typical of industrial farms. People often say how nice it is to get out to the farm and "back to nature," but this kind of corporate megafarm—the kind prevalent in the United States and, increasingly, everywhere else—feels anything but natural. It's cleared and leveled. The beetles, bugs, vines, and thickets of the natural world have been sprayed, treated, and extinguished with pesticides, insecticides, and fungicides, rendering miles of fields into sterile deserts that grow one type of crop, year after year, indefinitely.

Like Kray, I'm lamenting the vicious circle the photo portends. For a little while crops will flourish; then the land will suffer disruption in the water cycle from drought, loss of soil fertility, and reductions in flows of rivers and streams, among other ecosystem declines. Forests protect water resources, regulate local climate, and support biodiversity. The rich soil, which took many years to create and was anchored by the trees, will likely be eroded away. And a valuable carbon sink will be lost.

Kray has spent a lot of time talking to people about climate smart agriculture in Uruguay and other countries since 2013. He seemed reluctant to share his experiences in Paraguay. In India, among the forward-thinking villages, I felt hope. The land the Indian farmers were tilling was already degraded from years and years of harvests, but in good faith they were trying new practices that conserve organic matter and natural resources. Farmers in Rwanda and Senegal and Vietnam were doing the same, under different cropping systems. They've improved their soils with organic matter and no-till farming and are learning to grow new breeds of higher-producing beans on vertical supports to get more from each acre. In Vietnam farmers are adapting to planting breeds of rice that can tolerate higher salinity.[1] But in this stretch of Paraguay, and countless other parts of the world where vast deforestation is still occurring in the name of large, industrial-scale monocropping, the future looks bleak.

Finally, I got it out of Kray. "I think destroying ecosystems to grow cattle and to grow big monocrop soybeans and to grow palm oil, when there are already sufficiently degraded areas to put those crops on, is climate stupid," he said emphatically.[2]

When I returned to the office, I found a satellite image of the same location that shows a clear demarcation between Paraguay and Argentina. Argentina's forests are slightly better off than Paraguay's but probably not for long. Deforestation there continues, the result of expanding soybean farms, and is threatening the so-called cloud forest and the Chaco ecoregion, one of the largest forested biomes in South America, which houses plant and animal species found nowhere else on Earth.

Kray's tree, and the climate-stupid agriculture it represents, remained with me. I wondered if all deforestation—indeed, all methods of modern commercial agriculture—could best be described as climate stupid. The tree could have easily been in Brazil's Amazon or in Indonesia, where deforestation has accelerated dramatically in recent years, as land is cleared to produce palm oil for processed food and personal care products. The growing world demand for processed foods, which drives the need for vegetable shortening (soybean oil and palm oil, among others), threatens to destroy even more forests. About half the items on grocery shelves, from ice cream and chocolate to soaps and detergent, contain palm oil, a product that is also used as a biofuel and for cooking oil.[3]

The 3.1 trillion trees on this planet work as Earth's lungs. They absorb air. More precisely, they absorb the carbon dioxide that accumulates in the atmosphere, and as trees grow and get energy from the sun, the carbon remains in their branches, trunks, and roots, and the trees release the oxygen. Tropical rainforests act as Earth's thermostat, by regulating temperatures and weather patterns. As we lose the forests, we compound the loss by adding to the atmosphere (instead of subtracting) 12 to 15 percent of all greenhouse gas emissions—roughly the same amount as the world's planes, trains, and automobiles combined.

What I learned from Kray, and scientists who work on sustainable development and food products that have been certified as ecologically safe, is that the solutions exist. It's possible to go from stupid to smart. Often it is simply a matter of will. And critical for moving forward is understanding the big picture—where forests are most precious and most endangered. Of course, to raise environmental concerns about the development of rural areas is to risk appearing insensitive to economic needs. Many people living in these areas have led a hand-to-mouth existence for centuries, and modern farming offers them an opportunity to have a middle-class life. Kray's work seeks to balance these needs. But to ignore the potential damage is its own kind of risk—one from which we all might suffer.

I remembered Kray's photo again at the Council on Foreign Relations in Washington, D.C., in October 2015, when I attended a private briefing that explained how experts planned to advise Indonesia's president about how to reform his country's rogue foresting sector. During the summer and early fall a huge number of forest fires had broken out in Indonesia, caused by the burning of Indonesian forests and peat lands for agriculture. The strong El Niño had led to dry conditions, and between July and late October, Indonesia had suffered roughly 120,000 such fires, eliciting sharp protests from Singapore and other Southeast Asian nations fed up with breathing the noxious haze. Malaysian Prime Minister Najib Razak told *The Guardian* that Indonesia should convict plantation owners responsible for the harmful smoke.[4]

The fires have long outraged the international community, but at last *national* anger from within Indonesia had the potential to slow its runaway deforestation. But these fires certainly weren't the first. Similar blazes in Indonesia's rainforests in 1982 and 1983 had shocked both the locals and the world. The logging industry had embarked on a decades-long pillaging of the country's woodlands, opening up the canopy and drying out the carbon-rich peat soils. Preceded by an unusually long El Niño–related dry season, the fires lasted for months, sending vast clouds of smoke across Southeast Asia.[5]

Fifteen years later, in 1997 and 1998, a record El Niño

year coincided with continued massive land-use changes in Indonesia, including the wholesale draining of peat lands to plant oil palm and wood pulp plantations, which convert wood into pulp, paper, and fiberboard for furniture.[6] Large areas of Borneo, the giant island Indonesia shares with Malaysia and Brunei, and Sumatra burned, and again Southeast Asians choked on Indonesian smoke. In the ensuing years Indonesia's peat and forest fires became an annual summer occurrence. The pall of smoke that drifted over Southeast Asia in 2015 was only the most visible symptom of decades of disastrous policies in Indonesia's corrupt forestry and palm oil sectors. In recent years changes in the country's land use have made Indonesia the world's sixth-largest emitter of greenhouse gases, and the 2015 fires propelled it into the top five, behind China, the United States, the European Union, and India. On thirty-eight of fifty-six days during the peak burning, carbon emissions from Indonesia's peat and forest fires equaled the daily emissions of the entire U.S. economy.[7]

More than a half-million people sought medical help for acute respiratory illness. Air quality in cities in Sumatra and Kalimantan Province in Borneo measured about 2,000 on the Pollution Standard Index; a reading of 300 is considered hazardous. Six Borneo provinces declared a state of emergency, and some residents were even evacuated to ships to escape the smoke. At least nineteen people died, and thousands of premature deaths will likely result from extended exposure to the smoke, public health experts said.[8]

The haze has closed schools, grounded flights, and angered key Indonesian trading partners such as Malaysia, Singapore, and Thailand. Economic losses from damage to agriculture, forests, health, transportation, and tourism were estimated at $16 billion in Indonesia. The burning typically stops when seasonal rains arrive in October. Rain fell in Kalimantan and doused some of the fires in late October 2015, but the strong El Niño conditions delayed the autumn rains, which did not arrive until well into November. In addition, the drained peat lands in many areas were nearly fifty feet deep—when they smolder unabated, they burn much like underground coal fires.[9]

But will Indonesia's latest haze crisis prompt effective reforms? Certainly, pressure has been mounting. The issue was discussed again during the United Nations climate negotiations in Paris in December 2015. Indonesian President Joko Widodo reiterated his support for an absolute moratorium on new licenses to develop peat land. That moratorium has been in place since 2011 and has focused attention on egregious cases but has not been comprehensively monitored.

Widodo said that the government would review existing licenses and that in the meantime license holders would be prohibited from opening up any peat land that had not yet been drained and planted. Palm oil companies with significant land banks of undisturbed peat land would not be allowed to convert them to agriculture, saving up to 7 million acres. Widodo also instructed his minister of the environment and forests to initiate a program to restore disturbed peat land. "If implemented, these measures will stop the bleeding," said Frances Seymour, the former director general of the Center for International Forestry Research in Indonesia and now senior fellow at the Center for Global Development, a think tank in Washington, D.C.[10]

Jim Leape of the Woods Institute for the Environment at Stanford University said that the catastrophic fires, and the White House meeting between President Widodo and President Obama in late October 2015, were catalysts for change. Leape said that the palm oil sector could continue to expand—without destroying forests—by intensifying production on already-degraded lands. "The largest palm oil companies in Indonesia are committed to that task," Leape said. "You have many leading civil society groups poised to support such a shift. You have international partners poised to support such a shift."[11]

Widodo's announcement was a bold step, but it remains to be seen whether his actions will back it up. The licensing freeze will not be passed into law, prompting concern over its effectiveness, Seymour explained.[12] The forestry ministry is famously corrupt. Questions remain about whether Widodo has the authority to invalidate licenses that have already been issued. But because the fires have created such an intense health and economic crisis, perhaps this is a tipping point that offers the best chance for reform in generations.

Seymour said there is reason for optimism. "Following through on this agenda would be consistent with President Widodo's professed desires to prevent a recurrence of the public health emergency caused by the fires, to weed out corruption, and to support indigenous rights," Seymour said. "This time, the stars could be aligned for a land-use paradigm shift." Seymour said there is also much work to be done on the international level to drive home the value to other countries of protecting Indonesia's forests and peat land.[13]

The Paris climate negotiations helped devise solutions to Indonesia's deforestation crisis, as experts discussed what policies and independent monitoring tools could be deployed to reduce the country's emissions by 29 percent by 2030. One potential model is Brazil, which has dramatically reduced its deforestation rate through heightened law enforcement and government policies that encourage development of previously degraded lands, according to Seymour.[14]

🌐

In both Brazil and Indonesia nongovernmental organizations such as Greenpeace have persuaded international corporations such as McDonald's to stop buying palm oil, soybeans, or other agricultural products grown on deforested land. In Brazil the environmental community received a major boost when the country hosted the first Earth Summit in Rio de Janeiro in 1992. Seymour said the summit shined a spotlight on forest and environmental issues and helped nurture a community of civil society organizations, economists, and scientists who developed tools, such as satellite monitoring, to slow deforestation.[15]

"Brazilian policy makers have seen that the conservation and restoration of tropical forests is not only good for the environment, but also for quality of life, the economy, and foreign relations," said Solange Filoso, a research assistant professor at the University of Maryland Center for Environmental Science. She has worked with Brazilian researchers to monitor reforestation to reverse the loss of biodiversity and assist in the recovery of ecosystems services—work that could be applied to other developing countries with tropical forests.[16]

Brazil transitioned from dictatorship to democracy in 1985, a process that did not begin in Indonesia until 1998. But Widodo's announcement of an end to further conversion of peat land indicated that conservation groups and activists are gaining greater influence. "I think it is fair to say that Indonesia is, in certain ways, following Brazil's example, but with Indonesia's own special features," Seymour said.

Demand for Indonesia's palm oil is still growing, and the country plans to increase production. It is the world's largest producer of palm oil, supplying 80 percent of global demand, 33 million tons of palm oil in 2014. By 2020 Indonesia expects to produce 40 million tons.[17]

Palm oil is a deceptive commodity. It has many benefits. It produces more oil per acre than other crops. For instance, an acre of oil palms produces much more oil than does one acre of soybeans or sunflowers or rapeseed (canola) or other oil sources. This efficiency is part of the reason palm oil has become a coveted commodity. But conventional methods of production still encourage rampant deforestation, greenhouse gas emissions, and loss of biodiversity. The huge environmental downsides should make leaders think at least twice before sacrificing long-term planetary health for short-term economic gain.

The Roundtable on Sustainable Palm Oil is the world's largest organization to certify that a product was produced responsibly. Producers can earn the right to use the roundtable's green label to market their palm oil as certified sustainable. The roundtable does not permit the use of fires within certified plantations, except under extreme circumstances. With 3,000 members it has the support of global food producers and suppliers. Even the Environment Program of the United Nations signed an agreement with the roundtable as part of a broader effort to conserve global biodiversity. But the certifier has been criticized for, among other things, failing to require zero deforestation by its members and continuing a policy that allows some members to continue environmentally damaging practices. Several investigations have exposed deep flaws in its compliance standards.[18]

Membership in the roundtable consists of thirty-three nonprofit organizations as well as palm oil planters, refiners, and consumer goods manufacturers.[19] The roundtable reported in a statement that instances of fire in 2015 on the 137 certified palm plantations in Indonesia numbered in the single digits. Because the fires were so widespread, environmental groups read this statement in disbelief. One of the thirty-three nonprofit members resigned from the group in 2016 because of what the nonprofit called "the purposeful stifling of the RSPO by its members and administration, who for whatever reasons are increasingly allowing for it to be weakened by the day." Their hope for, and ultimate frustration with, the roundtable was no surprise.[20]

Indonesia supplies palm oil to all the world's major brands. Despite their pledges of sustainable business and corporate social responsibility practices, these brands have not been able to exert pressure to break the fire cycle. In its 2015 investigation Greenpeace reported that the roundtable "has struggled to break the link between palm oil and forest destruction" and connected some of its prominent members to "violations of Indonesia's forest moratorium, to deforestation, and peat land destruction and to the fires." In February 2016, the roundtable launched a new, stricter palm oil label in response to the fires. Called RSPO Next, the label aims to halt deforestation and fires.[21]

As of December 2015, Indonesia had fined twenty-three companies for contributing to the fires.[22] While the roundtable holds out hope that the businesses can insist on ethical palm oil production in their supply chains and stop ongoing deforestation, environmental groups have plenty of reason to worry. After environmental groups filed grievances against one of the largest conglomerates, IOI Group, a founding member of the roundtable, the organization waited six years before suspending the Malaysian company's membership—after three IOI subsidiaries were alleged to have broken some laws and failed to prevent rainforest destruction. The suspension from RSPO hurt IOI

Group's stock price on Malaysia's stock market, but the company's shrugging response to its loss of a sales premium for certified sustainable palm oil suggested that a market solution is not the answer. A representative of IOI Group said the demand for palm oil that has not been certified would continue unabated. "The only effect is that IOI will not be able to earn a [certified] premium on our oil which represents only a very small percentage (less than 0.5%) of our revenue," said a representative of IOI in a statement.[23]

Not all forests are created equal. Research has found that tropical forests play an important role in absorbing carbon, accounting for about 40 percent of the total land-based absorption.[24] One argument for the preservation of tropical forests is their ability to act as sponges that store vast amounts of carbon and help regulate the global climate. The reason to maintain forests in Indonesia is they have the potential to provide even greater economic gains. Forests that are managed well and sustainably help improve the resilience of farmers, preserve water availability, and improve food security.

In Indonesia proponents of a moratorium on further deforestation for palm oil say that the oil palm plantations are not producing efficiently and that bigger yields could be realized from land that is already deforested if plantation owners took simple measures to improve productivity. In addition, oil palm plantations could expand on land that is already degraded; this acreage typically has low levels of biodiversity, stores little carbon, and is not likely to store much carbon in the future. Indonesia, where land classified as *imperata*—grassland—has overrun logged forests and prevented reforestation, has millions of acres of such land. One approach would be for the oil palm plantations to expand into these areas, rather than the vital carbon-rich peat land, and allow palm oil derived from this acreage to be certified as sustainable.

In the last one hundred years Earth has lost as much tropical forest as it did in the previous 2,000 years. In Borneo about 75 percent of the lowland rainforest is gone. Since Jeffrey Hayward, a forestry specialist, studied Amazonian deforestation at the University of Washington in the 1980s, 15 percent of the Amazon rainforest has been lost, and more than a million more acres go every year.

Hayward has worked at Rainforest Alliance since 2000, and since then at least 370 million acres of forest have been lost globally. His organization was established in the mid-1980s to address the problem of deforestation, but the problem remains large. Rainforest Alliance has had impact. It has helped to transform business practices and does so by connecting consumers and markets to sustainable forest and food products, but much work remains.[25]

Hayward and his team have helped foster the responsible stewardship of millions of acres of farm and forest globally, and they've positively impacted the lives of 5 million people. By 2020 his organization plans to have 200 to 250 million acres under responsible management, he said. "We'd like to see that 10 million hectares [almost 25 million acres] of that land is climate smart, and we would like to be providing training for five million [people], and forest dependent communities. In order to do that it's going to take a hell of a lot of work."

Rainforest Alliance has worked with small palm oil producers in Indonesia as well, but most of its efforts are concentrated in the most vulnerable countries all over the world to try to balance forests, food security, and farming. Hayward spoke with me about three little-known areas with the potential to fight deforestation.

The first is technical assistance and training for using climate-smart agriculture in forest and land systems. Hayward's team helps people and institutions undertake constructive responses to climate change while also enhancing their agricultural productivity and food security. Hayward has worked to limit deforestation, reforest areas, and improve farming practices in Uganda, high on the hillsides of Mount Elgon (see chapter 2 for the discussion of deforestation that led to landslides in the foothills of Mount Elgon). Farmers, typically subsistence farmers, are subjected to periods of drought, insufficient rain, or too much rain that severely erodes their soils. They have coffee farms and also plant subsistence crops, but they have not received much technical assistance or investment in their production. Their coffee is of poor quality and has low yields. Rainforest Alliance teamed up with Taylors of Harrogate, a British tea company, and DFID, a UK aid agency, to bring more sustainable farming practices to 6,000 farms atop Mount Elgon and help them improve the productivity of their coffee.

First, Hayward established field schools for farmers. Then he trained lead farmers, and those lead farmers in turn trained others. The lead farmers also established model farms, clustered in communities throughout the ecosystem, that demonstrated best practices and were opened to visitors in an effort to disseminate information. One of the main lessons was soil conservation, including the use of mulch and drainage ditches to help control runoff. One technique is to concentrate and cluster the treatment of the coffee hulls that are removed when workers wash the berries. Now the coffee bean–processing area of a model farm has a composting system; the resulting fertilizer goes into the farms and increases the productivity of their coffee plants.

"We were able to work with them to better manage the water that's used in the washing of the coffee hulls and reduce water use significantly, and also to treat water waste considerably, which is important for health—health of people and health of streams," Hayward said. He continued:

> By introducing better practices of how they managed the coffee trees themselves, the coffee bushes, through pruning and proper fertilizer placement, and timing, we were able to help them increase their yield significantly, and also, in terms of restoring the overall forest landscape there was quite a number of trees planted through the project with nurseries set up. So that's a situation where we are improving people's lives directly through their mainstay, which is coffee production.[26]

Hayward has also worked in sustainable cocoa production in Ghana, in western Africa, where cocoa farms dominate the landscape. He said the farms were similar to those in Uganda. They were low producers, often yielding only half of their total capability. The farming techniques did not make use of fertilizers or efficiency with labor and capital, and were encroaching on the remaining forest. Climate-smart agriculture, for these farmers, really meant helping them through the same type of farmer field schools to develop better practices and introduce restoration of trees, better placement of trees that serve as shade for the cocoa, proper pruning practices for the cocoa, better fertilization through development of compost, and restoration of fallow areas. And, above all, the focus was understanding what farmers need economically and how to help them get there.

By working with 5,000 farmers in Ghana in an area of roughly 135 square miles, with the support of Olam, an international commodities trader, and the U.S. Agency for International Development, Hayward learned that farmers in the cocoa sector didn't want to limit their farming to cocoa. They wanted to expand into beekeeping, animal husbandry, and the raising of livestock. They needed training, and to do that they needed some seed money, which Rainforest Alliance also provided.

Further, the Uganda farmers needed more community organization. Hayward explained,

> This is one of the things that's so fundamental, and often we miss this case, is that if communities are going

> to be sustainable, they need to be communicating to each other, they need to be sharing lessons learned, they need to have the governance spaces in order to organize and coordinate, and through setting up clusters of co-operatives and landscape management boards, we're able to help them to devote a better system for actual overseeing how the wider landscape is managed.[27]

His third example came from Ecuador, in the Amazon, where the Huaorani peoples in Napo Province seemed to be ideal beneficiaries for a government reforestation program. The government wanted to try to restore deforested areas. The program had flaws, and Rainforest Alliance worked to address them. First, the original reforestation program resembled commercial forestry, with monocultures of fast-growing high-yield tree species. Hayward said that those tree species didn't grow well in the Amazon, and no one had any experience with them.

Second, people in this area were shifting to growing a new variety of plant called naranjilla, an herbaceous plant similar in size to a golden tomato or chunky eggplant bush that can grow to eight feet and produces a small orange fruit that's quite popular domestically. Said Hayward,

> If you're going to introduce a commercial reforestation plan into an agricultural community, you have to figure out a way to make the two mesh, and they weren't meshing at all, so what we did was work to figure out the kind of native species from the Amazon that the people

> are familiar with, that they like, and that they would use, and that would also enable them to grow this naranjilla.

It seemed like a difficult proposition at the time. But Hayward and colleagues at the Rainforest Alliance found a way to develop a mixed agroforestry system with native species and naranjilla. He then laid out a twenty-year management plan to move the growers from an introductory period of naranjillas and small trees to larger trees (which were selected by the local community, had multiple uses such as timber, fruit and forage, and were native to the location) and naranjillas and finally to just larger trees and then repeat, repeat, and repeat rather than using a reforesting plan that eventually would encroach on the neighboring Amazon. It worked.

Hayward and his colleagues at the Rainforest Alliance pore over the characteristics of suitable commodity crops and try to figure out local solutions to help people increase their agricultural productivity, revenue, and resiliency to climate change—all at the same time. "A lot of the things that our organization does, and a lot that we talk about needing to do about climate-smart agriculture, is nothing new," he said.

> These are techniques that have been around for thirty or forty years, many of them even longer. It's just a matter of, can you get them to the people that need them, can you get them to them in a way that fits with

> what they do locally, and can you adapt to new practices, and institute new forms of institutions that are needed at landscape level that will help people to take up best practices that are widely accepted and well recognized as actually solving problems?[28]

As these examples show, meeting the growing demand for palm oil in the future without further destruction of tropical forests, as well as producing other kinds of food the world will need, will require forest protections, a coordination of solutions, and enduring political will. Holger Kray has studied the projections repeatedly, and even the most optimistic projection, he said, points to our being able to produce only 20 percent more food, when we will likely need 35 percent more food by 2050. Additionally, with 25 percent of greenhouse gases coming from agriculture today, "we are far from other sectors, which have made big commitments in reducing emissions," Kray reminded me. He noted, for example, that the transportation sector has the electric car.

The consequences of not achieving the food production that human populations need will be grave, he said. "Roughly 40 percent of all arable soil in Africa will not be fertile any longer by 2050 if we continue treating soil like dirt. And if we do treat soil like dirt, we shouldn't be surprised we're not getting there with yields," he said. But in Uruguay, a coastal country neatly tucked away between Brazil and Argentina, Kray finds a countryside showcase that offers hope.

Uruguay has quadrupled its agriculture production since the turn of the century. It is among the top ten exporters of red meat, wool, and rice, and it does all that while dramatically reducing the per-unit carbon emissions from agriculture. Why is that? Since 2000 resilience to climate change has been the explicit focus of Uruguay's agriculture policy. "There is no subsidy that is not conditioned on enhancing your resilience, on improving the environment, on treating your soil better. It reminds me a little of the European Union, which we criticize so often, but more than half of its subsidies, at this point in time, [are] already conditioned on environmental compliancy all across our planet," said Kray.[29]

So what does Uruguay do? Every farmer has to submit soil management plans to the government. "I know saying that in the United States of America sounds like an alien concept, but, believe me, in Latin America it was an alien concept, too, when Uruguay started [the requirement]," he said.

> Farmers have to submit soil management plans, and these soil management plans are overlaid in an electronic database with soil quality maps, so if you compare the vulnerability of your soil, the soil quality, your exposure to erosion, your water content of your soil with the production plans of the farmer, then all of a sudden you might find out that certain parts are not meant to be monocropped with soy.

If this happens, Uruguay takes the liberty of keeping that farmer away from monocropping soil.

Basically, Uruguay's policy bolsters soil fertility. Soy exports have gone down a little. "But that is for a reason!" Kray emphasizes. "That's because Uruguay saves its soil."[30] An article in *The Independent* (U.K.) reported on research findings that if the United Kingdom continues its current farming practices, it will have roughly one hundred harvests remaining until the soil is entirely depleted.[31] In contrast, Uruguay will have added to its soil's fertility, Kray explained, "and [it] will still have quite a bit of fertility left, and I make this case because it is really a showcase."

What else makes Uruguay a showcase? In 2006 Uruguay started individual livestock registration using ear tags in cattle. The original motivation was public health concerns about hoof-and-mouth disease. Buyers wanted to track their food from farm to table. But early on, natural resource managers and policy makers understood that knowing where livestock is located is an incredible tool for measuring resilience to climate change. To better understand where a country is most vulnerable, Kray explained the value of having data on vegetation density, livestock density, and family farm registries, which can overlay with soil quality registries and with soil moisture management plans, for instance. "We do that easily from a satellite nowadays, and then in the wake of an incoming drought, you will know the most vulnerable spots in a country, the hot spots," Kray said. "That's exactly where you have the lowest vegetation, the highest livestock density, the highest incidence of family farming. These are more vulnerable than corporate farms, and so on, and so forth, and then the government can intervene for that very location, and doesn't have to put pressures, [or] other resources, into an entire county, state, or country."

As an economist Kray believes that investment incentives can support a farm sector and a food sector as they become more resilient to climate change. "By knowing where the hot spots are, you can deploy [the incentives] in a very targeted way," he said.

The little glimpses into a country, and into a system, that Kray provided me were truly stunning. Kray is quick to point out that Uruguay is one of several countries that have exemplary policies that help create incentives for an agriculture sector: "It's like Africa's Rwanda or even the European Union." These are places that cherish soil health, water in that soil, and, perhaps most important, biodiversity, by maintaining old seeds in seed banks that scientists can use to breed varieties that are more resilient to the looming challenges—"be it climate or be it us," Kray said.[32]

Rice

Knee-high weeds were growing in planter pots in a small room at a U.S. Department of Agriculture lab just outside Washington, D.C. Light, heat, and carbon dioxide reached the plants at steady levels. For more than a month the weeds survived in the same conditions expected to be Earth's norm about thirty years from now—carbon dioxide levels equivalent to an urban traffic jam and temperatures tipping into the dangerous zone for the planet's health.

You'd think any plant would choke, but these particular weeds—a wild plant called red rice—were thriving. Scientists in the United States are conducting intensive experiments to cross hardy weeds with food crops such as rice and wheat. Their goal is to make these staples more resilient as higher temperatures, drought, and elevated carbon dioxide levels pose new threats to the world's food supply, even as we welcome 2.6 billion more people to the global population.

Breeding allows scientists to build desirable traits into crops and limit their vulnerability to pests and disease. Modern plant breeding techniques run a spectrum, from the classical to the computer enabled. Classical breeding is fairly straightforward; it deliberately interbreeds, or crosses, closely related plants or wild relatives. For instance, a lima bean that is resistant to powdery mildew may be crossed with a high-yielding but disease-susceptible bean, with the goal of introducing mildew resistance without losing productivity. In this chapter I will discuss some research that shows how new breeding solutions might help plants cope with the environmental stresses expected as a result of climate change.

Lewis Ziska, a plant physiologist with the USDA's Agricultural Research Service, is primarily a weed man. Weeds beguile Ziska. They may be the largest single limitation to global crop yield, yet they also have traits that are useful to plant growth. Red rice, which is actually a wild grass and likes to compete with cultivated rice, can adapt to more carbon dioxide and heat by producing more stems and grain—red rice has 80 to 90 percent more seed than cultivated rice, which is grown specifically for human consumption.

Plant breeders and plant physiologists now are capitalizing on those traits and counting on all sources of genetic variation, including the weedy lines of rice, to improve productivity in cultivated crop varieties.

Plant physiologists like Ziska usually define weeds in two ways: "An unwanted or undesired plant species" and "early vegetation following soil disturbance." Ziska thinks a third definition could be more suitable: the unloved flower. "A weed is a plant whose virtues have yet to be discovered," he said, paraphrasing Emerson. Why weeds? When other plants are wilting at extremes of temperature and rainfall, weeds thrive. Ziska studies weeds for their redemptive qualities. His research in this area was bolstered on a sweltering day in an abandoned industrial lot in Baltimore, twenty-five miles from his USDA research center, where he observed weeds that were two to four times bigger than weeds growing on his rural test plot. The urban weeds prompted further research on weeds that could be valuable to raising crops in high-carbon, high-heat scenarios.

Ziska tries not to call red rice a weed. Sometimes he calls it skanky rice, but mostly he refers to it as wild or feral rice. All crop plants were wild at some point. They became domesticated in the same way cows and pigs became domesticated on farms: through breeding and selection. Wild and feral crop relatives are the original source of raw genetic material from which all modern crop varieties were first developed, but these reservoirs of natural variation have not been well studied. Ziska believes weeds and heirloom varieties will be a key part of the solution to crop survival in a warming world. "The feral cousins of today's crops may allow us to adapt to meet food security needs," he said. "This paradox of weeds I find fascinating. Let's turn lemons into lemonade."[1]

Wheat breeding has already made significant progress in producing weedy lines; this has led to the cultivation of edible wheat from those lines. Because it has a large genome, wheat can, with relative ease, incorporate traits that better withstand heat and drought. Matthew Reynolds, head of the wheat physiology program at the International Maize and Wheat Improvement Center in Mexico City, says that wheat producers are fortunate because they are slightly ahead of the curve in developing heat- and drought-resistant varieties.[2]

All this hard work has already paid off. Cross-breeding with wild relatives of crops to make them more resistant to pests, diseases, and environmental threats began in 1977 and by 1997 had led to an annual benefit to the global economy of about $115 billion, primarily through increased production, according to a study by a Cornell University plant geneticist.[3]

Consider again the weed called red rice. As the name implies, it looks a lot like cultivated rice—the staple food for more than 3 billion people in the world. This wild Asian grass can grow unnoticed alongside cultivated rice for a long time. And, boy, does it grow. It also propagates quickly; because red rice is taller than other rice plants when fully mature, its seeds, spread by the wind, go farther than those of cultivated rice.

Herbicides cannot be used to control red rice because it

is too closely related to most cultivated rice. Herbicides would kill the cultivated rice too. Once red rice becomes established in a field, it is so aggressive that it will cut a field's rice yield by 80 percent. Within five years it can become the dominant species. And while technically edible, it's almost impossible to harvest, because its seeds fall to the ground and shatter. The goal for researchers has been to transfer to cultivated rice the traits that make red rice so hardy.

With a growing global population, the demand for rice and other cereals is expected to rise by 14 percent every decade. But climate change is expected to cut into some of those crop yields. Today's high temperatures stress the rice plants, limit growth, and shorten their growing seasons. The carbon dioxide accumulating in the atmosphere also causes weeds to outpace crop growth. Coastal deltas are major rice-growing areas worldwide, and repeated coastal flooding is becoming more frequent as a result of rising sea levels and intensifying storms. Vietnam, one of the top exporters of rice globally, is already losing land in the Mekong Delta. In other places, such as Indonesia, land suitable for growing rice attracts the developers of oil palm plantations.

According to the International Rice Research Institute, the environmental stresses that hamper a rice harvest affect about 30 percent of the 700 million poor in Asia who live in rainfed rice-growing areas. Breeders are developing rice varieties that can withstand conditions in the future, such as more frequent and intense problems of drought, flood, heat, and cold, and soil problems like high salt and iron toxicity.

Although nearly 2,000 international gene banks hold seeds of wild crops and ancient landraces (a local cultivar that has been improved and adapted to their environments through traditional breeding), "they are not used to their full potential in plant breeding," said Susan McCouch, professor of plant breeding and genetics at Cornell University and head of the Susan R. McCouch Rice Genetics Lab. "There are still vast reserves of valuable genes and traits hidden in low-performing wild ancestors and long forgotten early farmer varieties that can be coaxed out of these ancient plants by crossing them with higher-yielding modern relatives. These crosses give rise to families of offspring that carry a myriad of new possibilities for the future."[4]

McCouch said the plant breeder's job is to apply a combination of insight, field experience, technology, and innovative breeding strategies to select the most promising offspring and prepare them for release as new varieties. The breeding system allows ancient traits and genes to be constantly recycled and recombined, giving rise to an infinite range of new possibilities with every generation.

"The same process happens in nature," she said, "but the plant breeder can bring together parents from diverse sources that would never have found each other in the wild." Weeds are unlikely heroes of biodiversity that could benefit modern, cultivated plant varieties in the face of climate change, McCouch said. She's counting on these diverse ancestors to possess natural traits that can help rice and other food crops be more resilient to climate variations.

To survive in the highly variable environment of the future, crops will need to sense and respond to their individual environments. "That's a lot different than breeding a plant for conditions we know it will encounter, when conditions are fixed," McCouch said.[5]

For example, researchers found a trait in an ancient variety of wild rice that senses when the plant is underwater and allows it to tolerate being submerged while still growing, McCouch said. Researchers identified the gene and then bred that trait into a line of rice now being used in Bangladesh. While most rice varieties can withstand only three or four days of being submerged—leading to annual losses of $1 billion that disproportionately affect the poorest farmers in the world—this rice can virtually hold its breath for two weeks and survive oxygen starvation, a necessary attribute for surviving the unpredictable rainfall and more frequent flooding expected with climate change.[6]

The Submergence 1 gene, as it is called, has already saved thousands of acres of rice in South Asia from catastrophe since it was introduced in 2009, according to the rice institute. Flash floods or longer-term stagnant flooding can affect rice crops at any stage of growth. But when flooding occurs during the crop's first six to eight weeks after planting, it can wipe out the crop; this is of particular concern in places like Bangladesh and India, where flooding already costs farmers up to 4 million tons of rice per year—enough to feed 30 million people.[7]

In the future researchers will continue to add traits to core varieties of rice, "much like you put ornaments on a tree," McCouch said, but the breeding time could accelerate with better information. When more becomes known about the characteristics of all the wild and ancient varieties of seeds in gene banks, future breeding programs will have the ability to move hundreds of genes at a time from wild plants to modern cultivars—a game-changing concept, she said.

To access the necessary information about the 7 million samples of crop varieties and wild relatives in the world's gene banks, researchers are translating the descriptions and names from other languages, organizing each gene bank, and putting the data in a master directory. McCouch is working on an international effort by sixty-nine organizations, called DivSeek, to identify traits, genes, or genomic composition contained by the plant material in each gene bank.

Ultimately dozens of crops could be improved by reintroducing genetic variations from wild ancestors, particularly in countries hardest hit by climate change, McCouch said. In East and West Africa, as well as Latin America, breeders are looking for new sources of genes to boost drought tolerance in traditional varieties of corn, and they are turning to ancient varieties of millet and sorghum to replace corn in regions where drought is too prolonged and soils are too poor to grow corn. In Africa and Latin America, where cassava is a staple food among the poor, breeders are seeking genes from all kinds of exotic species to help cassava become more disease and pest resistant and increase its physiological resistance to postharvest deterioration, McCouch said.[8]

Landraces and wild ancestors of beans likely possess traits that will help modern crops mature faster, for example, earlier in the season, and give them increased natural tolerance of pests, disease, and extremes in temperature. Breeders in Latin America and Africa are looking to landraces and wild varieties of peanuts not only for new sources of heat tolerance but also for resistance to the human carcinogen aflatoxin, a class of toxic compounds produced by some molds that are more likely to contaminate crops under hot conditions.

Traits originally identified in wild relatives or landraces will find their way into food production in a few different ways, according to McCouch. Submergence tolerance in rice, for example, can be introduced through cross-breeding and marker-assisted selection, which is a combination of traditional genetic breeding and molecular biology that has become a valuable tool in selecting organisms for traits of interest, such as disease resistance. Traits with more complexity may be introduced from wild relatives and landraces using traditional crossing and a more complicated form of marker-assisted selection that can predict which individual plants in a breeding program are the most valuable parents of the next generation of offspring. This approach, known as genome-wide prediction, is cutting edge and is just beginning to be used in public breeding programs, although it is more common in the private sector, particularly for corn.

When wild ancestors are used as parents in crossing and selection, the process is essentially that of traditional breeding, assisted by the use of molecular markers, but it requires no special biosafety testing or labeling because it is not a genetically modified organism. The two are entirely different beasts. Instead of spending decades of physically identifying plants that will have the desired qualities, such as increased yield, ability to stand up to drought, and more, breeders can speed up the process with this DNA screening. Marker-assisted selection minimizes the "wait and see" period in traditional breeding. "This will continue to be a common use of the materials that DivSeek is helping to document and make more accessible," McCouch said.[9]

In a process still under development, a breeder may choose genome-editing technology to change the sequence of a particular gene in a modern variety to match the gene from a wild species in order to confer a trait from the wild plant, such as disease resistance. Another option is cloning a gene from a wild species and introducing it in a modern crop using transgenic technology (biotechnology), McCouch said, but that approach would require years of biosafety testing.

Ultimately global warming may prove to be an unexpected stimulus for a huge wave of advancements in plant breeding that haven't been needed until now, McCouch said. “Climate change is introducing completely new, unpredictable weather patterns in which we have rain during the dry season and drought during the rainy season and flood season,” she said. “So, under climate change, we are breeding with a completely new concept in mind.”[10]

Agricultural science must move quickly to keep pace with climate change. Plant breeding takes time, but adaptations are needed now. As evidence of the existence and effects of climate change piles up, plant science is in a race to help food crops surmount the risks for food security.

Improved rice breeding has a long way to go. It takes about ten years for a crop to go from breeding to production and another five years to bring it to distribution to farmers—a painstakingly slow process of selecting populations of offspring that contain combinations of traits and genes that have never been used in agriculture before and then testing their resiliency to environmental stresses.

And crop resilience does not come in the form of a silver bullet. “It will not be one new trait, one super crop variety or one new management system that allows us to meet the world’s demand for food,” McCouch said. “It takes time, effort and training to make significant genetic progress when utilizing these ancient sources of variation.”[11]

⊕

Plant breeding has been making the headlines in recent years because of widespread misconceptions about the role of genetic modification in farming and the resultant “genetically modified organisms,” better known as GMOs.

Well-intentioned efforts to be transparent about what is in the food we eat have, unfortunately, become conflated with scares about food safety. In late July 2016 President Barack Obama signed into law a measure requiring the labeling of foods that contain certain GMOs, and it gave the U.S. Department of Agriculture two years to write the rules governing the labeling.[12] Those labels are likely to fail to explain what modifications were made and why, which can make a world of difference. Certain potatoes grown in the Andes of South America, for instance, have been bred using traits from wild South American potatoes to tolerate frost, protecting the food security of hundreds of thousands of people in low-income communities. In Hawaii a type of papaya was bred to tolerate a virus that threatened to destroy the crop. And in the Philippines a variety of eggplant was modified to resist an increasingly devastating infestation by a fruit and shoot borer, reducing the use of chemical pesticides that also harm the environment and affect the health of farmers. Understanding GMOs is a lesson in how to think critically, and the main point is that each case of GMO use should be considered on its merits. Generalizations about GMOs ignore the complexities of their benefits, including sustainability and affordability versus environmental and health concerns.

Genetic modifications of a wide number of plants are possible, but the biggest users of GMO technology have been large-scale growers of agricultural commodity crops. In the United States canola, corn, cotton, sugar beets, and soybeans are the main genetically engineered crops that are planted; upward of 90 percent of these crops are modified. These foods produced with GMO have not been found to be harmful to people who eat them. The most common method of making crops resistant to insects is adding a gene from a naturally occurring bacterium in soil, *Bacillus thuringiensis (Bt)*, which acts as a toxin that protects plants from pests. This genetic modification has reduced the need to spray fields with insecticides. Other genetic modifications, such as the addition of a gene that allows the plant to survive when the field is sprayed with weed-killer chemicals, have caused more concern. Now more weed killer, or herbicide, is needed to control weeds.

While consumers' increasing desire to know what is in their food is a positive trend overall, the blanket condemnation of GMOs coming from some circles is misguided. Indeed, improved breeding and genetic modification are key tools in the fight against global hunger and climate change. On balance, there's no such thing as a perfect breed or best variety because agricultural crops—both GMO and conventional seeds—will need to evolve to adapt to their growing conditions amid climate change and a variety of pests and disease. Yet blanket praise for GMO crops is misguided too. We know that global food security comes from supporting biodiversity. As multinational corporations increasingly consolidate and focus on developing a handful of commodity crops, farmers have less real choice in the diversity of seeds commercially available to them at an affordable price. Agricultural crops will need to constantly change and achieve greater diversity in their breeding to be able to withstand a variety of pests, weeds, viruses, and conditions in the future, but with so few companies controlling the market for seeds—mainly soybeans and corn—many worry food security is at risk. The issue of GMOs is a complicated one. New breeds of plants—whether they are conventional or tinkered with genetically—need to be considered case by case while understanding how they can be used to improve food security, health, and the environment without posing threats to humans and the environment. The call for more transparency around GMO labeling may open the door to further transparency of the food system, which could be a very good thing for a growing planet. Consumers could gain much more critical information about the food they eat, including how much water was used, what chemicals were used and why, whether antibiotics were used and which ones, animal welfare considerations at feedlots and slaughterhouses, and even how the conditions for farm workers are becoming more socially responsible.

PLANET

Hot Fish

Along the shores of Lake Victoria in the Nyanza Province in Kenya, a thirty-five-year-old widow who is a mother of six works as a fish seller. She cleans and dries fish in the hot sun, then sells them at the local market. Like most women in her situation, she relies on the whims of fishermen to give her the best from their daily catch. To ensure that she has access, she uses both money and her body. The "sex for fish" system is known in the local Luo language as *jaboya.* And as the supply of fish dwindles the practice is becoming more and more common in communities surrounding the lake.

Every year growing numbers of people in search of a better life migrate from the increasingly eroded and degraded outlands to Lake Victoria and its watershed. It's now a densely populated rural area and will only get more crowded as conditions become drier and hotter. These migrations have brought new burdens. Pollution from waste, damage to the vast and delicate wetlands from so many humans working and living nearby, and interference with fish breeding sites have severely diminished the number of fish the men land in their nets. That makes men's jobs harder, as men own most of the fishing boats, and a woman's job almost impossible. "When you are a woman and you want to get into the business of selling fish, you must be ready to lose your pride and use your body for bargaining," the widow told an interviewer. "Being ready to give sex as and when it is needed by the fishermen . . . it guarantees your survival here on the beach."[1]

More people than ever rely on fisheries—and aquaculture—for a source of food and income. Lake Victoria is a painfully clear example of how crowding, pollution, social justice, and competition for natural resources converge to splinter human health and food security. And while fish have a growing role in feeding the world, because cattle and other livestock are too environmentally costly, fish are no magic answer. Pressures on fisheries are immense. The latest estimates find that fish now account for about 17 percent of global population intake of animal protein; in some coastal and island countries, it can top 70 percent.[2]

The heat and hunger of our planet are also putting intense pressure on many of the world's lakes and oceans, and the stress will only increase over the next twenty years. Global warming and rapid population growth are contributing to the loss in the diversity of fish species. How we address the survival of lake fish, marine species, and aquaculture will determine a great deal about the future quality of life and health as well as the economies of communities around us. How much we can gain by improving the resilience of our fisheries and the conservation of critical ecosystems that support marine life—and the risks if we fail to do so—is the focus of this chapter.

But it won't be easy. Places like Lake Victoria, where fish are declining, also tend to be places where the incidence of human rights abuses and food insecurity is high. Let's go back to the sex-for-fish dilemma and what it means in a global context.

Bordered by Kenya, Uganda, and Tanzania, Lake Victoria is the second-largest freshwater source in the world. It's a biodiversity hot spot, which means the region is home to a significant number of species threatened with destruction by human activities. In fact, it's one of twenty-five biologically rich areas around the world identified as having lost at least 70 percent of its original habitat. Lake Victoria is also an important fishery and a water source for agriculture. As climate change makes the prospects of nearby farmers bleaker, more people will look to the lake for food. Harvested fish supply more than 22 million people locally with a substantial portion of their protein.[3] The value of the catch at local markets, known as landed value, was $590 million in 2009, with $250 million of this total earned from the export of Nile perch, according to research by the NGO Lake Victoria Fisheries Organization. Regional population growth, which is among the highest in Africa, and economic development have led to declining water quality, reduced fish stocks, and industrial pollution.[4]

It's also an important case study in how the fight to empower women can also greatly improve health and food security. The practice of *jaboya* is not new, but with dwindling fish supplies and high levels of HIV infection, it is clearly unsustainable. In Kenya's Nyanza Province HIV prevalence is 14.9 percent, twice the national average of 7.4 percent.[5] (On the other side of Lake Victoria, in Uganda near the Tanzanian border, 43 percent of the men in a fishing community live with HIV.[6])

The majority of women living on the shores of Lake Victoria have either a direct or indirect role in the fish trade, according to the Ministry of Fisheries in Kenya. The widow involved in the practice of sex for fish said she was aware she was vulnerable to contracting HIV but was willing to risk it to provide for her family. "You know you can get HIV . . . but then you remember you have a family that needs to be provided for, and you say, let me die providing for them," she told the interviewer.[7] Because 90 percent of the women were widowed when their husbands died from AIDS, and the women continue to live in poverty, they remain vulnerable to sex for fish.

"Fish trade that goes along with sex for fish continues to be one of the greatest challenges in the prevention of HIV in Nyanza. . . . There are still challenges which involve the economic and social vulnerabilities of the women involved in the trade," said Charles Oka, the AIDS and sexually transmitted infections coordinator for Nyanza Province, in an interview with *IRIN/PlusNews*.[8] The female fish sellers who choose to not have sex with the men often do not get access to the catch of the day or cannot buy enough fish to feed their families, even if they have the money.[9]

A handful of organizations are trying to combat the health risks, poverty, and food insecurity of these women. The groups want to revive the lake and its fish by integrating solutions, such as the environmental and health programs discussed in chapter 5, but the local people have to address so many problems in their daily lives that protecting the environment and natural resources, like fisheries, has not emerged as a sustained priority.[10]

So a group of NGOs came up with a plan to tackle all these problems at once. The Peace Corps donated six fishing boats to women's groups in Nyanza Province. Now some women are beginning to break the cycle of vulnerability. They are eliminating their dependence on fishermen by having the men work for them. "When you have nothing, those who have something must tell you to bend over backwards for them. Now we have boats and we will no longer be at anybody's mercy," Millicent Onyango, one of the beneficiaries of the "No Sex for Fish" project, told *IRIN/PlusNews*.[11]

The economic empowerment of women will be key to ending this dangerous fish-for-sex trade. By owning a boat, they no longer have to play along with *jaboya*. While six boats might seem small, such initiatives could spell the end of *jaboya* if replicated over time.

🌐

The practice of *jaboya* encapsulates the interdependence of human systems and natural systems. Ecosystem health is important for sustaining human health and food security, and vice versa. Whether tropical lake fish can adapt to increased temperatures, for instance, is another looming threat. Using the population, health, and environment model to find solutions to the problem may offer one way out.

Clive Mutunga, a population, environment, and development technical adviser at the U.S. Agency for International Development, said that since the mid-1960s the Lake Victoria Basin has suffered substantial environmental degradation. Declines in the number of fish caught, poor water quality, deforestation, and loss of biodiversity in the lake and its wetlands have eroded environmental health. People in this region are some of the poorest in the world, and as such their day-to-day needs for survival mean they place little importance on protecting the lake. The drivers of the degradation, which include agriculture, pollution, deforestation, overfishing, and increasing industrialization, have been addressed through environmental regulations, but these are not enforced.[12]

Tropical lakes with warm water may seem like idyllic havens to catch the abundant fish that live and grow there year round, but the reality is sometimes quite the opposite. Some of the world's important lakes have provided early warning signs of the imminent challenges facing tropical lake systems. Consider Lake Bam in Burkina Faso and, on a larger scale, Lake Chad, once one of the largest bodies of water in Africa; the latter is only 10 percent of its size in the 1960s.[13] Julian Cribb writes about Lake Bam, part of the Volta River system in West Africa, in his book *The Coming Famine*:

> Its flow fluctuates with the seasons, expanding when the rains replenish it and contracting as the fierce heat of summer drinks up to two-thirds of its shallow waters. Of late, however, the rains have been sparse and the lake has dwindled even more severely in summer, fragmenting into a chain of muddy ponds less than a meter deep. Soil eroded by farming and grazing in the surrounding catchment is filling it in, too much water has been taken from its feeder streams, and the local climate appears to be drying out. The waters once supported nearly a thousand hectares of irrigated crops, but these have dwindled to one hundred and fifty. The nets of fishermen often come up empty. Yet the main sources of income in this, one of the world's most impoverished regions, are crops and fish.[14]

Climate change is rapidly warming the world's lakes. Tropical lakes have evolved over time with fairly constant temperatures, and a warming of 0.93 degrees Fahrenheit

(0.52 degrees Celsius) per decade is a sizable swing from the norm. That is a particularly big problem for Lake Victoria and the African Great Lakes, whose fish are a major source of food. Using more than twenty-five years of historical satellite temperature data and ground measurements of 235 lakes on six continents, a breakthrough study in 2015 found that lakes are warming an average of 0.61 degrees Fahrenheit (0.34 degrees Celsius) each decade, a rate that is greater than that seen in the oceans or the atmosphere. "We want to be careful that we don't dismiss some of these [tropical lakes with] lower rates of change," said Stephanie Hampton, of Washington State University, who published a paper on threats to freshwater ecosystems. "In warmer lakes, those temperature changes can be really important. They can be just as important as a higher rate of change in a cooler lake. The pervasive and rapid warming observed here signals the urgent need to incorporate climate impacts into vulnerability assessments and adaptation efforts for lakes."[15]

Around Lake Victoria, fishing increases food security, contributes to the economy, and adds to the stability of communities by adding tax revenues.[16] The future efforts to improve social systems, such as finding a way to end the sex-for-fish practices and advancing knowledge about conserving natural systems, could hinge on the very knowledge that fish ecologists are now pursuing in a global research project about the physiological ecology of freshwater fish in tropical areas.

Michael Cooperman, a fish ecologist with Conservation

International, along with researchers from Carleton University and McGill University in Canada, have been examining the genetic capacity of fish living in tropical inland freshwater systems to tolerate changes in temperature. Called the Hot Fish project, their study examines select species of fish in three distinct areas: Tonle Sap, a lake in Cambodia; Lake Victoria in Uganda; and the Amazon in Manaus, Brazil. Tropical freshwater lakes and inland rivers are at greater risk for a decline in species than more temperate lakes, even though temperate lakes are likely to experience a larger swing in temperature. That's because tropical species have not evolved to handle variance; in fact, the temperature resilience of the kind of freshwater tropical fish humans rely on for food is not known. Cooperman's study is one of the first to monitor how future climate change—specifically, a slow increase in temperature—will affect fish growth, reproduction, and survival.

"When you think about life in the temperate world, the species of life that live here have to be able to tolerate changing temperatures. You have seasons. You go from winter to summer and back to winter. Not too long ago, in an evolutionary sense, we had an ice age!" Cooperman told me.

> Organisms have to be able to go through a range of temperatures. They have a certain genetic capacity to tolerate this. If you are a trout in a Montana stream, you are going from winter temperatures around freezing to

> summer temperatures in the low seventies. Bass and sunfish do the same thing. But if you think about the tropics, particularly the lowland tropics, temperatures are stable year in and year out. There is no seasonality. So there was never any driving mechanism to make the organisms that live there tolerate changing temperatures."[17]

In other words, if you are a fish and live in an environment with a stable temperature, you are at risk from a small change in temperature—even a 4- to 6-degree Fahrenheit (or about a 2-degree Celsius) increase from climate change.

In Cambodia 75 percent of the animal protein that people eat comes from freshwater fish. Nobody knows how those fish are going to respond when their waters warm up, Cooperman explained. In Cambodia his team tested two species of fish, but over 200 species exist in the lake. In Uganda and the African Great Lakes region, freshwater fish also are important in the diet, particularly that of poor people, but the Nile perch (native to the Nile and Congo river basins but introduced into Lake Victoria) is also a major export. "The Nile perch fishery of Lake Victoria is more important from the employment standpoint and the tax revenue standpoint than it is [from] a food security standpoint," Cooperman explained. Most of the Nile perch are exported to Europe. The Hot Fish study tested Nile perch along with two other species often harvested by locals.

The study used a technique called respirometry to

measure how much energy the fish were using at various temperatures. Cooperman put it this way:

> Picture yourself sitting in a chair doing absolutely nothing. You are completely at rest. But your body is using a certain amount of energy just to stay alive. Your heart needs to beat. Your diaphragm needs to expand so you can breathe. We hope your brain is working to keep your fundamental processes going. That energy requires oxygen because that is how your cells make the oxygen, through respiration. You breathe in oxygen, and when you exhale there is a little less oxygen and more carbon dioxide. That is how your cells get oxygen. Now let's say you stood up and you did jumping jacks for five minutes or ran in place—doing some form of exercise, you are using more energy. You breathe in and out faster and deeper, and the amount of oxygen is greatly reduced. When you breathe out you use a lot more oxygen. So that when you were at rest, with no exercising, that is basal metabolic, or resting metabolic rate. When exercising, your metabolic rate goes up. Maybe five minutes of jumping jacks isn't enough to get you to your passing-out ability. But if we did it for three hours nonstop, at some point you are going to hit that maximum where you pass out and cannot go on for any longer. Your body cannot produce any more energy. That is your maximum rate. That is the same for fish or any other critter in the world.[18]

For cold-blooded organisms like fish, the amount of energy they are producing, which is a proxy for the amount of energy they are using, is temperature specific. If a trout is in 50-degree (Fahrenheit; 10-degree Celsius) water and not swimming—not fishing a current, not actively producing gametes in preparation of spawning, just hanging out in the dead of winter—that is the equivalent of basal metabolic rate, Cooperman said. At the other end of the spectrum is its maximum rate, or the equivalent of a person's pass-out rate. The scope (maximum minus minimum) represents a fish's discretionary energy budget. When the temperature changes, so does the scope—as 50 degrees Fahrenheit warms above 53 degrees Fahrenheit, the trout has to spend just a little more energy to stay alive. Its basal metabolic rate will increase because it is moving away from an optimum temperature. But at the same time maybe its processes—its enzymes or muscle efficiencies—are a little more or less effective. The maximum rate may change as well. This continues up to a point. "If you think about Darwin, and take basic Darwinian evolution," Cooperman said, "an organism is optimized to the environment to which it evolved."

Consider the trey riel, known as money fish, a silvery fish in the lake Tonle Sap in Cambodia. When its discretionary energy budget is maximized, the trey riel has energy for growing, reproducing, finding food, and fighting its enemies. As temperatures change, this capacity is reduced. Fish are just like people in that they are strategic in how they use their energy. "It could be they have less growth. It could be that they produce less eggs or sperm. Or the individual eggs or sperm have less energy density in them," Cooperman said. "It could be that there is less energy for behaviors like foraging or fighting your enemies or whatever it is."

The hot fish research is among hundreds of studies to examine, among other things, how a shift in management practices could save fish-dependent poor people from food insecurity, malnutrition, and health risks. One global analysis in 2016 examined fish availability and concluded that more than 10 percent of the global population could face micronutrient and fatty-acid deficiencies driven by fish declines—in lakes, marine systems, and aquaculture—in the coming decades, especially in developing countries along the Equator.[19]

In the spring of 2016, I met with members of that fisheries and food security research team. They were working at the National Socio-Environmental Synthesis Center at the University of Maryland, where I am a senior fellow. The team was examining a hypothetical perfect storm in a world that won't be able to fish its way to feeding billions more people by midcentury.

The research team was comprised of ecologists, public health scientists, marine scientists, and environmental affairs researchers. Led by Christopher Golden, a research scientist at the Harvard T. H. Chan School of Public Health, and Dave McCauley, an assistant professor at the University of California, Santa Barbara, the researchers examined new databases on global fish catch and human dietary nutrition. They found that the vulnerability of poor fish-dependent populations in the tropics has actually been *underestimated* and that these are the very places where fish resources are under the most intense pressure.

At its heart the problem is a simple one of supply and demand, the researchers explained. Global fish catches peaked in 1996, while Earth's human population is expected to rise through 2050, to top 9 billion. But those figures oversimplify a problem also affected by natural processes, economic pressures, international regulations, and plain old human needs. Golden said that it is important to include human nutrition, along with the preservation of biodiversity and economic considerations, in determining how fisheries are managed. The work of the synthesis science team estimates that, in the coming decades, 11 percent of the global population—845 million people—will be vulnerable to micronutrient deficiencies because of their reliance on seafood, a figure that climbs to 19 percent, or 1.39 billion people, if the researchers count only micronutrients found in animal sources, such as vitamin B_{12} and DHA omega-3 fatty acids.[20]

While fish are recognized as an important source of protein, they also provide often overlooked micronutrients—like vitamin B_{12}, iron, and zinc, Golden said. According to the report, micronutrient deficiencies can affect maternal and child mortality, cause cognitive defects, and negatively affect immune function. About 45 percent of mortality in children younger than five worldwide is attributable to malnutrition.

The problems facing subsistence fishing populations, such as those around Cambodia's Lake Tonle Sap, or Lake Victoria in Africa, are not solely the result of overfishing, which has been successfully addressed in some locations through sound management. Both destructive fishing practices and coastal pollution are degrading the aquatic environment before climate change is even factored in. In the oceans acidification bleaches coral reefs, while rising temperatures force tropical species poleward. The effects of climate change could reduce catch by 6 percent globally and by as much as 30 percent in certain tropical regions. Warming tropical seas will hold less oxygen and cause fish to get smaller, cutting overall biomass by about 20 percent by 2050.

People who live in industrialized nations can compensate for the nutritional gap left by a decline of fish, Golden noted. But those in developing nations often have few alternatives.[21]

Even among developing nations, however, there is much variation in the threats to local fish supplies. Madagascar, a large island nation where Golden has worked on the interface of human health and the environment for seventeen years, suffers most from unsustainable fishing practices and foreign fleets in its waters—issues that could be addressed with better management (as well as the public health interventions discussed in chapter 5). Small island developing states, like Kiribati, the Maldives, Palau, and Vanuatu, however, may have more intractable problems. Climate change is likely to push local fish species away, while rising seas may flood low-lying areas where aquaculture is practiced. And most people in these islands are too poor to afford high-quality replacement foods or supplements.

In low-income, developing nations, such as Madagascar, people will fall back on foods with fewer nutrients, such as rice and tubers. Those in wealthier nations, like the Philippines, Brazil, Mexico, and Indonesia, meanwhile, may buy inexpensive processed foods to replace fish, which would increase their populations' risk of metabolic diseases rather than malnutrition, Golden said:

> Wealthy nations are somewhat immune to these environmental effects. They can create systems of food imports, intensive agricultural food production, fortified foods, and supplements that buffer them from the potential pitfalls or consequences, whereas it is poorer populations dependent on the direct pathway from the environment to their own well-being that are most at risk. There's almost a reverse Robin Hood system where the wealthier nations are now going into biodiversity-rich areas, with robust fish populations, and using foreign fleets to capture resources—both legally and illegally—and bring them back to wealthier populations that don't need them.[22]

Some see aquaculture, the farming of fish and shellfish (and seaweed) in water environments, as the answer. In fact, in 2014 for the first time more people ate fish from aquaculture than from wild-caught sources. Farming fish is an efficient way to convert feed into animal protein. Many people believe that it relieves pressures on ocean fisheries, but research is finding the potential benefits of aquaculture are not quite as cut and dried as they seem. As practiced in oceans, marine aquaculture has downsides. It can contribute to localized water pollution, and the fish food itself can contain disproportionately large amounts of wild fish. As well, the nutritional needs of fish raised in aquaculture vary by species. Catfish and tilapia are herbivorous and eat feed that is made from a mixture of plant proteins found in corn and soy, vegetable oils, minerals, and vitamins. Carnivorous fish like shrimp and salmon eat other fish, and when farmed they consume more fish than they produce. Some estimates say that as many as two pounds of wild fish feed are needed to produce one pound of farmed fish. (Land use and energy use also are at a premium to supply food, tanks, and ponds for fish like tilapia and catfish.) On balance the farmed fish do add to the world food supply, but the cost to oceans and land is not insignificant. In addition to their outsized consumption of wild-caught feeder fish, farmed fish live in pens that can add to water pollution, and their habitats can add to the destruction of mangroves and wetlands (in the case of farmed shrimp).[23]

One aquaculture company, Verlasso, has found ways to reduce the amount of wild fish in feed for farmed fish and adopt more ecologically sound management practices such as raising fish in clean, cool waters away from industrialized or sensitive areas and providing more space in each pen for fish to grow and develop. Others are beginning to follow the lead of Verlasso, a joint venture of DuPont and AquaChile, an aquaculture operation based in Patagonia.[24] While global aquaculture production has exploded, much of the production is intended for tables in the developed world or for developing nations' urban elite. In addition, aquaculture is not entirely divorced from wild fisheries, given the content of the feed.[25]

While it's unlikely that wild harvests will provide the same nutrition for the significantly larger human population at midcentury, better management can improve catches by as much as 10 percent, Golden said, and, if done with an eye toward human nutrition, can avoid hundreds of millions of cases of malnutrition. "The hopeful thing is that policy and management [have] been shown to rebound fisheries on the scale of a decade," Golden said. "So it's a really important time to be sounding this alarm so nutrition-sensitive policies can be implemented."[26]

In its study of land use, the World Resources Institute analyzed the potential for aquaculture to more than double production by 2050 without exceeding the global limits of fishmeal and fish oil supplies and found:

> The catch of small fish from "industrial" fisheries used for fishmeal is on the decline, dropping by half from 30 mega tons in 1994 to 15 mega tons in 2010. Aquaculture has still grown, in part by diverting more and more of the fishmeal and oil from livestock feeds. But aquacul-

ture now consumes 63 percent of global fishmeal and 81 percent of fish oil and therefore there is little left to divert from other uses.[27]

Experts agree that for aquaculture to meet projected need levels in 2050, the industry will have to continually improve feeds and feeding practices to further reduce fish-based ingredients.

⊕

One solution used to produce farmed salmon features omega-3 oil extracted from genetically modified yeast. This lab-made omega-3 oil can replace part of the wild-caught fish that salmon eat, and its use has been making the case for sustainable fish farming. It was the brainchild of Scott Nichols, who started his devotion to seafood at a young age.

As a child in Hawaii, Nichols woke early. By 6 a.m. the five-year-old had usually reeled in his first catch from a predawn fishing expedition with his dad. The first time the boy carried his bounty—a whole fish—into the family kitchen, his father gave him an impromptu lesson in gutting and cleaning fish, and a lifelong love affair with seafood was born.

Much later in life Nichols found himself in the lucky position of being able to turn that love for fish into a business venture. A biochemist with a doctoral degree from the University of California, Los Angeles, who also studied business at Wharton, Nichols was in charge of business development for DuPont. It was 2006, and the company had just had a breakthrough with genetically modified yeast: it found it could be used as a substitute for fish oil because it supplies the desirable omega-3 fatty acids previously found only in animals. But Nichols knew it was a breakthrough in another area as well. "In the blink of an eye, I realized that we could solve a big problem with salmon aquaculture," said Nichols, who founded Verlasso.

In 2006 salmon aquaculture was consuming about 80 percent of the world's fish oil, and the market for farmed fish was still growing at a rate of 8 to 10 percent per year. Oily fish like anchovies, menhaden, and mackerel provide the main source of fish oils, and their harvests have been threatened as their populations become depleted. Supplying feed for fish farming of salmon is partly responsible for the depletion of the world's oily fish, but other factors have been at work, including the increase in consumer demand for omega-3 fatty acid vitamin supplements and overfishing.

"We are looking at a future where there would be no more fish oil to be had," Nichols told me. "I thought, if we are able to provide the omega-3 to the salmon using DuPont's yeast that is rich in omega-3s, and use far fewer wild-caught feeder fish for the diet, it would do a lot of good for the oceans while sustainably supplying farmed salmon with omega-3s."

Verlasso's method of salmon aquaculture reduces reliance on wild-caught fish by an astounding 75 percent. Four pounds of wild-caught feeder fish are typically needed to produce the fish oil to make one pound of salmon. Verlasso, on the other hand, relies on just one pound of wild-caught fish to produce one pound of salmon. "We have lowered the fish in/fish out ratio. One in and one out," Nichols said. Verlasso also identified ways to get down to using about three-quarters of a pound of wild-caught fish per one pound of fish produced, and Nichols believes the company will be able to achieve that over time. Although the joint venture's current focus is on raising Atlantic salmon, the feed could be useful for other salmon species, such as steelhead or coho, Nichols said.[28]

"Everyone recognizes we can't continue to harvest forage fish to feed oil to salmon," he said. "Some people ask, How do we use these [forage] fish with [the] most efficiency? The proper question is, how do we use them not at all? They need to be food sources themselves." According to Verlasso, the company saved more than 6 million pounds of feeder fish in 2013 by significantly reducing the amount of fish oil it uses in raising salmon. However, Nichols said that providing omega-3s to the fish through the yeast is more expensive than using fish oil.

"The company formulates the fish diets based on optimal performance rather than least cost, and a number of the ingredients in our feed are more expensive than those used in traditional salmon aquaculture," he said. The fish grow in the southern Pacific Ocean off Chile, reaching harvest size in about two years in pens with fewer than four salmon per ton of water, or about 50 percent more room per fish than the industry standard.

While fish sustainability is Nichols's business, he said

he often thinks of the pressing problems of world agriculture and how to produce enough food to feed the expanding global population in decades to come. "We've got to find ways to do more with less. How do we develop agriculture practices that operate in harmony with the environment and allow us greater intensity? There is a lot of ocean out there that we could use for raising plants and animals," Nichols said. "We need to find ways to use the oceans properly to raise more food. I heard a great quote from former NOAA [National Oceanic and Atmospheric Administration] administrator Jane Lubchenco: 'It's OK to use the oceans, just not OK to use them up.' "[29]

He points to a 2012 study from the Food and Agriculture Organization of the United Nations that reported 87 percent of the world's fisheries were harvested at or above their sustainable limits. In 2014 that number rose to 91 percent.[30]

"We are not going in the right direction," Nichols said.

> I hope it is axiomatic to say that it is indefensible to harvest a fishery above its sustainable level. A thornier question is how we should respond to roughly half of the world's fisheries' being harvested at the upper limits of sustainability. Operating at the very edge leaves little or no room to accommodate things unforeseen. . . . There seems to be precious little international enthusiasm to talk about how to reduce pressure on fisheries, but it is surely needed.

If people are going to continue to eat fish, he said, they must be farmed and raised sustainably. "All agricultural production, whether on land or in water, has environmental effects," Nichols said. "The key consideration is that we manage those effects so that our practices today do not impinge on our ability to practice in the future."[31]

Based on Verlasso's continuous improvement in its award-winning aquaculture practices, the company's ocean-raised farmed Atlantic salmon was the first to receive the "good alternative" ranking in 2013 from the Monterey Bay Aquarium's Seafood Watch Program. Other seafood aquaculture companies such as Blue Circle Foods in Norway have also had recent success in gaining that rating from Seafood Watch. Looking ahead, Nichols said the next phase of work is a bit unglamorous but important: Verlasso is reexamining its farming practices to see how the company can raise fish more efficiently and more healthfully without increasing its costs. Verlasso hopes to achieve this while searching for ways to reduce the use of wild-caught fish because, he said, "it is imperative we diminish our pressure on wild fisheries."

After securing recognition from the Seafood Watch Program and setting a sustainable path for Verlasso, Nichols left the company in 2016 and started his own seafood consulting business to help others in the aquaculture industry improve their practices and their relationships with environmental nonprofits. He believes the seafood industry would benefit from the expertise of nongovernmental organizations—such as the Aquaculture Stewardship Council—in devising clear, understandable metrics that help meet sustainability requirements for responsible aquaculture. As a consultant, he is looking to bridge the chasm between sustainability and business practices, as well as accelerate how companies can achieve sustainability. Ultimately, Nichols said, producers of aquaculture and sustainability-minded NGOs want the same thing: to nurture the ocean while also maintaining a healthy ecosystem.[32]

Small and Sustainable

The sky across the prairie in northern Iowa is massive, and the green fields stretch as far as the eye can see. Laid out on a grid, they are especially mesmerizing when the wind blows in gusts, as it does this day, sending gentle waves across the deep green sea of soybeans. In many places wind turbines, often more than 200 in one direction, add a surprisingly modern twist to the bucolic scene.

Down the gravel road off Highway 71, south of Carroll, Iowa, is a 240-acre plot where the Sibbel family has been farming since 1919. On the left side of the lane leading to the farm was a field of corn, grown from seeds that were not genetically modified. When harvested in the fall, the grain, corncobs, and stalks from this field would feed the Sibbels' pigs and cows for the year. On the other side of the lane were soybean fields, the yields from which would be sold to the local farmers' cooperative for feed, and a grassy fourteen-acre pasture where a Black Angus bull and sixteen cows were grazing with their calves.

When I arrived, Scott Sibbel was standing on the back of a tractor and banging the side of a steel grain bin. He was trying to assess how much feed was inside the bin. Ten barns and sheds of varying sizes, in varying stages of disrepair, dot the land around his farmhouse. Many barns are red-painted originals. The first animals I met were a drift of piglets, five weeks old with six different mothers. They ran together like cousins at a summer picnic. They were startled by my presence at the door and playfully darted down the aisle of the barn and out to a fenced yard that contained a feed trough and water. Each day Sibbel changes their straw bedding, which keeps the area clean and the pigs healthy. At nearly thirty pounds, they were almost a week away from getting weaned off their mother's milk and moved to another barn.

Sibbel, thirty-five, was among the younger farmers to take up the trade in his area when he started in 2005. He raises chickens, pigs, and cows in a humane way and sells the livestock to Niman Ranch, a processor and distributor of high-quality meats from animals that are raised under strict sustainable practices by a national network of independent farmers and ranchers. Sibbel is part of a growing movement among U.S. farmers to produce food efficiently by working with nature rather than against it. When he first learned about Niman Ranch, the company had one hundred livestock producers in its network. Now 765 farmers raise animals for Niman. Not only do these farmers take care of their land in a sustainable way, but they also allow the animals to live outdoors and interact with their litter mates. This is the approach Sibbel learned from his father but a far cry from the huge monoculture commodity farms across the United States and the predominance in his county of intensive, feed-lot cattle and pig confinement operations, where thousands of animals spend their whole lives in tight, indoor spaces. Sustainable farming operations like Sibbel's meet the requirements of diversity, resilience, and robustness that will be critical for American farms in a hot, hungry future.

Sibbel's transition to full-time farming wasn't easy. When he started, corn prices were low and a farm future looked bleak. "There was a negative vibe out there against young farmers who wanted to come back to the farm. My father and uncles were not encouraging. They tried to talk me out of it," he said. "But all I ever wanted to do was farm."[1]

After growing up on a farm and graduating high school, he left home to earn a degree in drafting and worked as a carpenter into his early twenties. Despite encouragement from his dad to stick to a day job, Sibbel was bent on earning his living as a farmer. While he worked at a drafting job, Martha, his wife, was in law school. One day she brought home a case study about Niman Ranch that she had read in one of her agriculture law classes.[2] She handed it to Scott. As he learned more about Niman Ranch, he saw his future. Right then he realized it was his route back to farming. Small farming could be profitable if he was selling to a niche market by raising animals much as his father had raised livestock. Scott also knew where he could start. His grandmother had recently moved away from her farm, and he could rent her land and her house. Just like that, his dream became reality.

Using the methods he learned as a kid and enhancing them with practices he learned from Niman Ranch guidelines, he began raising crops and livestock in an integrated system, which was the way he had always imagined doing it—with animals free to roam with their young, kept on clean bedding, and bred and raised naturally, without growth hormones or antibiotics, from good-natured boars and bulls. The animals have an all-vegetarian diet with no meat or bone meal. Sibbel follows specific guidelines for space and bedding requirements when his sows are farrowing, that is, giving birth to and nursing their piglets. Unlike the practices used for crated sows at confinement or factory farms, Sibbel's sows, and the sows of other farmers who raise pigs under Niman Ranch protocols, can move freely in their stalls to satisfy their instinct to build a nest when they give birth. Sometimes this bigger space means that sows will inadvertently lie on one or two of their piglets during lactation, but that is the gamble. Throughout the year Sibbel breeds and raises 500 to 600 hogs that weigh as much as 290 pounds by the time they reach Niman.

After a few years of successfully raising pigs, Sibbel also started raising Black Angus cattle for beef. His cattle are born in the spring and take about eighteen months to mature. He keeps the calves with their mothers in a grassy pasture for the first six months until they are weaned. The animals drink from a natural stream that flows through the pasture. After they separate from their mothers, the calves live in another pasture for a year before they return to Sibbel's barn, where they fatten up until they reach a market weight of 700 pounds or more. Each year he typically raises fifty to seventy-five cattle on roughly sixty acres of grassland pasture.

As a small family farmer under contract to Niman, he earns a higher price for his pigs and cows because he is compensated for the extra costs of raising the animals in a humane manner. The market is booming for this higher-quality, better-tasting meat; and if demand is any indication, it is worth the extra price.[3] Consumers can buy his pork and beef though specialty food markets, such as Whole Foods, and you will see it labeled as Niman's on restaurant menus.

Conservation farming runs deep at the Sibbel farmstead, and stewardship of the environment has always been at the core of its cropland success. Scott's grandfather built terraces on hilly parts of his fields in the 1930s and 1940s to keep the soil from washing away. "A lot of people plow right across the contours," Scott Sibbel said. Following the contours saves the topsoil from washing away and eliminates the sedimentation that can build up in rivers and streams from runoff. Sibbel has maintained his grandfather's terraces and has taken soil conservation measures even further by not tilling the land after harvesting, which adds more organic matter to the soil, and by planting cover crops.

In the late 1990s, when farm commodity prices were low, family farmers left the Midwest en masse, taking with them many sustainable conservation farming practices. Farms in places like Iowa consolidated, with larger corporate farms buying out the smaller producers. Farm sizes grew and the number of individual farms fell. Monoculture crop practices became the norm. Many farmers now plant corn on top of corn every year without rotating it with a soybean crop, an agronomic practice that requires the addition of large amounts of chemical nitrogen to the soil rather than allowing the nitrogen-fixing properties of soybeans to contribute it naturally. Livestock, too, became largely feed-lot operations. Big became a matter of survival—earning a living was a matter of scale. This was supported by public policy that tolerated harm to the environment and failed to reward practices that bolstered the resilience farming operations need in the face of climate change.

When Scott Sibbel was born in 1981, Iowa had roughly 125,000 farms and the average farm size was 275 acres—comparable to the current Sibbel spread.[4] By 2013, however, cropland in Iowa and in the rest of the country had become so consolidated that the number of farms fell overall and the size of farms practically doubled. The change in farm size has been even greater in Carroll County near Sibbel's place. Of the 1,000 farms that stretch across the county, more than a quarter of them are bigger than 500 acres and 10 percent have more than 1,000 acres.[5]

The foundation of Sibbel's operation is good breeding, not vast acreage. The sows at his farm have been bred to purebred Berkshire, Hampshire, Hereford, or Chester White boars that Sibbel bought from lines of premium breeders in his state. The result is offspring that have strong heterosis, or the tendency of a litter to show qualities of health and hardiness that are superior to both parents'. High-quality, well-bred pigs have fewer risks of health problems.

He has two barns where he farrows the sows, and once the piglets are weaned they move to other barns as they grow. He sells his full-grown hogs when they are about six months old.

He rotates the locations of his crops from year to year, and they generally include a mix of corn, soybeans, alfalfa, and winter rye, which is used as a cover crop to improve soil fertility. His soil gets an extra boost when he allows the cattle to forage on field residue after harvest and when he spreads the fields with manure from his livestock operation. The corn he grows is not altered genetically. While a modified seed might produce more grain per acre and hold up against pests, for Sibbel the decision not to grow with genetically modified (GM) seeds is one of both economics and preference. Typically the GM corn grown in his area is modified in three ways: to be resistant to drought, to be able to tolerate the weed killer Roundup, and to contain a toxin that prevents root worms and corn borers from ruining the crop. But the modifications don't protect a crop from all problems, and the seeds are pricey, he said. A bag of non-GM corn seed costs him $140 and will plant 2.5 acres, whereas GM seed costs $260 to cover the same amount of land. His cattle prefer non-GM cornstalks, which are not as tough, he said.

The benefits of small, sustainable farms like Sibbel's are many. First, the environment is better off when soils and waterways are healthy. Biodiversity in small farms is also greater, because crop rotation, cover crops, and a manageable animal production operation contribute to better-quality crops and livestock. Smaller farms also mean that main streets in local communities remain vibrant, unlike the ghost towns that often result when huge corporate farming rules an area.

Sibbel is a religious man, and he is committed to sustainable farming practices because he believes it is his calling. He doesn't like to judge how other people farm, but it's difficult for him to compare his own practices with the growth of huge industrialized farms—and even huge family farms—in his area.

After we visited a ten-acre pasture he rented for grazing twelve of his pregnant heifers and a bull some distance away, we drove past one of the large cattle operations in the county. It's a family-owned farm, but the similarities appear to end there. On this day, on a hillside of eighty acres, I could observe from the roadside thousands of cattle, confined in pens and standing at feeding troughs. Where

Sibbel grazed his cattle on roughly one acre per animal, this so-called family farm appeared to confine easily more than one hundred cattle to an acre. And it was not the only animal confinement operation around. As we drove through the county, I saw them everywhere, huge confinement barns for hogs and for cattle. I almost didn't believe what I was seeing.

The environmental side effects of such operations are immense. Because the animals in such systems receive growth hormones and antibiotics, the runoff and leaching of their waste into creeks, streams, and groundwater contaminates soil and drinking water. The manure lagoon of this particular operation appeared to be one of the larger bodies of water in the county. On the day of my visit a tall pipe was spraying the lagoon water onto a nearby field. The ammonia in the air was stifling.

Despite the sustainability benefits provided by small to medium-sized farms, subsidies for the industrial producers abound. Although Sibbel grows most of the grain and hay he feeds his animals, and raises them in pastures, he receives little or no support from federal programs except for a floor price during a drought when yields are poor and for an incentive to grow cover crops that improve soil fertility. Industrial livestock producers usually buy their grain, and feed grain subsidies have kept their total costs of production unusually low. Unfortunately, national farm policy favors the industrial producers, not the conservation farmers seeking a more sustainable path.

The ostensible reason for producing monocrops of corn and soybeans and raising livestock in industrial confinement systems is to produce the maximum amount of food to feed as many people as possible. Whether it does that is another question. The food grown and wasted by large-scale operations is immense. Plus, 47 percent of the corn crop in Iowa doesn't get made into food. It becomes ethanol and is mixed with gasoline for transportation fuel.[6]

One lesson that seemed apparent after touring the Sibbel farm is that small and midsized growers and producers in Iowa, and around the world, have a greater chance of achieving a sustainable operation because their model favors environmental protection, provides a more resilient economic and social fabric to rural life, and supplies nutritious food while minimizing overconsumption and waste. The Sibbel farm illuminates what the research scholar Tim Searchinger of Princeton University and the World Resources Institute calls "the great balancing act" on the "menu of solutions" that he proposes in *Creating a Sustainable Food Future*. He writes, "Each solution contributes to—or at least does not undermine—economic and social development and environmental protection."[7] I would argue that small farms that produce sustainable food can be a tool for safeguarding human well-being by reducing overconsumption of food, especially meat. That is, a portion of quality meat from the Sibbel Farm is unlikely to resemble a supersized portion from beef produced in a feed lot because it's more expensive and thus more likely to be consumed as a side dish. This reduces environmental impacts, particularly greenhouse gas emissions, as well as health implications and makes a strong case for reducing the overconsumption of beef. Doing so is also likely to reduce waste, because consumers in industrialized countries account for roughly half of the food loss and waste, and because consumers are less likely to waste an expensive piece of premium quality meat. Because Sibbel's farm is a less intensive operation, conservation farming protects soils and ecosystems and avoids overuse and pollution of freshwater. Finally, Sibbel's example shows that improvements in agricultural practices have a cascade effect. With the right combination of will, policies that support sustainable farming, and changes in consumer preferences for sustainably produced food, we can farm smarter and farm better for a sustainable food future.

Not long after visiting Sibbel's farm, I attended an invitation-only sustainable foods event that brings together scientists, thought leaders, and the media to learn from each other and about the coming trends. The panels of experts I heard over the two-day meeting covered the main elements of what it means to produce food sustainably, and they provided nuanced understanding of everything from world aquaculture trends to California's drought to climate change impacts on food security, and even the abusive practices on workers in the fishing industry in Southeast Asia. All kinds of people raised their hands to ask the expert panelists questions on food sustainability, from journalists and chefs to people running big food service companies and NGOs. As I was about to ask a question, one of the speakers said a word that cut the air.

The word was *safe*. Around the world, no matter where you live, the food you buy is expected to be safe. That's just expected, right? You don't go into the grocery store and pause each time you grab a bunch of bananas or a block of cheese and ask yourself, is this safe to eat? The point one speaker was trying to make was that just as people now expect their food to be safe—people in the future will expect their food to be sustainable. Policies now put limits on fishing where species are endangered, and future policies are moving toward getting traceability for all at-risk species in the supply chain. In the future under climate change—on a hotter planet with a lot more people on it—sustainable food will have a much greater meaning. Just as you now expect not to get sick from eating food you buy, the future the speaker envisioned was one where people expect their food to be raised without damaging the environment, to protect human health, to safeguard animal welfare, and to support local communities.

The speaker's example focused on the sustainability of seafood, but by the time the century reaches its midpoint, it is likely we will expect all food to be produced sustainably. The panel's discussion moved toward the great likelihood that all the food we eat will come with greater emphasis on its sustainability, as improvements in traceability and our knowledge of how water, nutrients, energy, methane, and carbon factor into our food choices. Other thought leaders attending the panels later told me that human factors, such as those around the social responsibility of farmworkers and food processors, will factor into a wider definition of "fair-trade"—and become another component to food sustainability. Further, the humane treatment of animals and the sustainable treatment of farmland and soils, such as I observed on the Sibbel farm, will one day become a reality in the future with the stronger, sustainable food system we need. There will surely be many solutions that will attempt to address the relationships between global warming and global hunger, more social innovations and new components in this never-ending cycle of smarter adaptations and innovations in the food system. As a growing planet constantly seeking to thrive, we will live with and adapt to the massive transformations that climate change will bring.

Notes

1: Introduction: The Fight to Close the Food Gap

1. "2015 World Population Data Sheet," Population Reference Bureau, August 2015, http://www.prb.org/pdf15/2015-world-population-data-sheet_eng.pdf.
2. Mario Pezzini, "An Emerging Middle Class," *OECD Yearbook 2012*, http://www.oecdobserver.org/news/fullstory.php/aid/3681/An_emerging_middle_class.html#sthash.9FKOEKCl.dpuf.
3. Michael Sherman, "The Co-op Movement," Vermont Historical Society, n.d., http://vermonthistory.org/research/research-resources-online/green-mountain-chronicles/the-co-op-movement-1919.
4. U.S. Department of Agriculture, "2012 Census of Agriculture—State Data," 2013, http://www.agcensus.usda.gov/Publications/2012/Full_Report/Volume_1,_Chapter_2_US_State_Level/st99_2_043_043.pdf (accessed January 20, 2016).
5. Steven McFadden, "Unraveling the CSA Number Conundrum," *The Call of the Land* (blog) January 9, 2012, https://thecallofthe land.wordpress.com/2012/01/09/unraveling-the-csa-number-conundrum/.
6. Verena Seufert, Navin Ramankutty, and Jonathan A. Foley, "Comparing the Yields of Organic and Conventional Agriculture,"

Nature 485 (May 10, 2012): 229–32, http://www.nature.com/nature/journal/v485/n7397/full/nature11069.html.

7. Sarah DeWeerdt, "Is Local Food Better?" *World Watch Magazine*, May 1, 2013, http://www.worldwatch.org/node/6064.
8. Steven L. Hopp and Joan Dye Gussow, "Comment on 'Food-Miles and the Relative Climate Impacts of Food Choices in the United States,'" *Environmental Science & Technology*, 2009, 3982-983, http://sciencepolicy.colorado.edu/students/envs_4800/weber_2009.pdf.
9. David B. Lobell and Sharon M. Gourdji, "The Influence of Climate Changes on Global Crop Productivity," *Plant Physiology* 160, no. 4 (December 2012), http://www.plantphysiol.org/content/160/4/1686.full.
10. S. Naresh Kumar, P. K. Aggarwal, D. N. Swaroopa Rani, R. Saxena, N. Chauhan, and S. Jain, "Vulnerability of Wheat Production to Climate Change in India," *Clim Res* 59 (2014): 173–187.
11. Timothy Searchinger, interview by author, South Royalton, Vermont, June 2, 2013. See also "Global Database on Child Growth and Malnutrition: Child Growth Indicators and Their Interpretation," World Health Organization, 2012, http://www.who.int/nutgrowthdb/about/introduction/en/index2.html.
12. National Agriculture Statistics Service, Jackson County, Minnesota, 1941–2015 data, *USDA/NASS QuickStats Ad-hoc Query Tool*, n.d., http://quickstats.nass.usda.gov/results/57B71333-2625-38A6-9378-968FB91C617A (accessed January 20, 2016); Seth Naeve, Iowa and Minnesota Soybean Yields, 1975–2005, graph in "Where'd All Our Beans Go?" *Crop Science* 41 (November–December 2001), http://www.extension.umn.edu/agriculture/ag-professionals/cpm/2004/SNaeveNov23.pdf.
13. United Nations Food and Agriculture Organization, "Reducing Poverty and Hunger: The Critical Role of Financing for Food, Agriculture and Rural Development," paper prepared for the International Conference on Financing for Development, Monterrey, Mexico, March 2002, http://www.fao.org/docrep/003/y6265e/y6265e03.htm.
14. "Gender Issues in Agriculture Labor," World Bank, http://siteresources.worldbank.org/INTGENAGRLIVSOUBOOK/Resources/Module8.pdf.

2: Women's Work

1. Republic of Uganda, "Table 2.2.1: The Most and Least Populated Districts in Uganda by Population Size," *National Population and Housing Census 2014: Provisional Results*, November 2014, http://www.ubos.org/onlinefiles/uploads/ubos/NPHC/NPHC%20 2014%20PROVISIONAL%20RESULTS%20REPORT.pdf (accessed January 19, 2016).
2. UNICEF, "Table 9: Child Protection," in *State of the World's Children, 2013* (New York: United Nations Children's Fund, May 2013), http://www.unicef.org/sowc2013/files/Table_9_Stat_Tables_SWCR2013_ENGLISH.pdf (accessed January 20, 2016).
3. Food and Agriculture Organization of the United Nations, *The State of Food Insecurity in the World* (Rome: FAO, 2011), http://www.fao.org/docrep/014/i2330e/i2330e.pdf, (accessed January 20, 2016). Republic of Uganda, *National Population and Housing Census 2014.*
4. Nyesige Safira, telephone interview by author, November 12, 2015. Musheshe Mwalimu, Susan Warshauer, Julia Pettengill, interviews by author, Cambridge, Massachusetts, April 14, 2015.
5. Catherine Namwezi, telephone interview by author, November 17, 2015.
6. Namwezi interview.
7. Namwezi interview.
8. "Supporting ISU," International School of Uganda, 2010, http://www.isu.ac.ug/page.cfm?p=8 (accessed January 20, 2016).
9. See, for example, R. C. Lesthaeghe, C. Vanderhoeft, S. Becker, and M. Kibet, "Individual and Contextual Effects of Education on Proximate Fertility Determinants and on Life-time Fertility in Kenya," in *The Collection and Analysis of Community Data*, ed. J. B. Casterline (Voorburg, The Netherlands: International Statistical Institute, 1985), 31–63; A. Jain and M. Nag, "Importance of Female Primary Education for Fertility Reduction in India," *Economic and Political Weekly* 21, no. 36 (1986): 1602–1608; S. J. Jejeebhoy, *Women's Education, Autonomy, and Reproductive Behaviour: Experience from Developing Countries* (Oxford: Clarendon, 1995); T. C. Martin, "Women's Education and Fertility: Results from 26 Demographic and Health Surveys," *Studies in Family Planning* 26, no. 4 (1995): 187–202; I. Diamond, M. Newby, and S. Varle, "Female Education and Fertility: Examining the Links,"

23–45, in *Critical Perspectives on Schooling and Fertility in the Developing World*, ed. J. Bledsoe, J. Casterline, J. Johnson-Kuhn, and J. Haaga (Washington, DC: National Academy of Sciences Press, 1999); J. Bongaarts, "Completing the Fertility Transition in the Developing World: The Role of Educational Differences and Fertility Preferences," *Population Studies* 57, no. 3 (2003): 321–336; S. Rutstein, "Fertility Levels, Trends and Differentials: 1995–1999," in *Demographic and Health Survey Comparative Reports No. 3* (Calverton, MD: Macro International, 2003).

10. United Nations Educational, Scientific and Cultural Organization, *Education Counts: Towards the Millennium Development Goals* (Paris: UNESCO, 2010), http://unesdoc.unesco.org/images/0019/001902/190214e.pdf.
11. Jere Behrman et al., "What Determines Post-School Skills? Impact of Pre-School, School Years and Post-School Experiences in Guatemala," cited in *World Development Report 2007: Development and the Next Generation* (Washington, DC: World Bank, 2006): 147.
12. George Psacharopoulos and Harry Anthony Patrinos, "Returns to Investment in Education: A Further Update," *Education Economics* 12, no. 2 (2004): 111–134.
13. Women and Population Division, Sustainable Development Department, Food and Agriculture Organization of the United Nations, "Women and Sustainable Food Security," in *Women: The Key to Food Security* (Rome: FAO, n.d.), http://www.fao.org/docrep/x0171e/x0171e02.htm; U.S. Department of Agriculture, "Farm Demographics—U.S. Farmers by Gender, Age, Race, Ethnicity, and More," Census of Agriculture: 2012 Census Highlights, table 3, http://www.agcensus.usda.gov/Publications/2012/Online_Resources/Highlights/Farm_Demographics/.
14. Cheryl Doss, "Killer Factcheck: 'Women Own 2 Percent of Land'=Not True. What Do We Really Know about Women and Land?" *From Poverty to Power* (blog), March 21, 2014, https://oxfamblogs.org/fp2p/killer-factcheck-women-own-2-of-land-not-true-what-do-we-really-know-about-women-and-land/.
15. United Nations Environment Program, United Nations Entity for Gender Equality and the Empowerment of Women, United Nations Peacebuilding Support Office, and United Nations Development Program, *Women and Natural Resources: Unlocking the Peacebuilding Potential* (Nairobi and New York: UNEP, UNEGE, UNPSO, and UNDP, November 2013), http://postconflict.unep.ch

/publications/UNEP_UN-Women_PBSO_UNDP_gender_NRM_peacebuilding_report.pdf.

16. Calestous Juma, "University for Women Key to Africa Agriculture," *New Vision*, May 12, 2014, http://www.newvision.co.ug/new_vision/news/1340575/university-women-key-africa-agriculture#sthash.TEGurQY0.dpuf.
17. Mwalimu Musheshe, lecture at MIT's Sloan School of Management, May 15, 2015, Boston.
18. New Vision Reporter, "ARU Holds Graduation," *New Vision*, November 24, 2015, http://www.newvision.co.ug/new_vision/news/1412080/aru-holds-graduation.
19. National Research Council, *Colleges of Agriculture at the Land Grant Universities: A Profile* (Washington, DC: National Academies Press, 1995).
20. Patricia Seybold, "Evolution of African Rural University," Customers.com, April 11, 2013, http://www.customers.com/articles/evolution-of-african-rural-university/.
21. Patricia Seybold, telephone interview by author, March 6, 2015.
22. Musheshe lecture.
23. Chip Heath and Dan Heath, *Switch: How to Change Things When Change Is Hard* (New York: Broadway Books, 2010), p. 45.
24. United Nations Population Fund, *The Power of 18 Billion: Adolescents, Youth and the Transformation of the Future: State of the World 2014* (New York: UNFPA, 2014), https://www.unfpa.org/sites/default/files/pub-pdf/EN-SWOP14-Report_FINAL-web.pdf.
25. "Large Landslide in Uganda," NASA Earth Observatory, March 11, 2010, http://earthobservatory.nasa.gov/NaturalHazards/view.php?id=43130.
26. David Mafabi and Tabu Batagira, "Landslides Bury Five Villages in Bududa," *Daily Monitor*, August 11, 2013, http://www.monitor.co.ug/News/National/Landslides-bury-five-villages-in-Bududa/-/688334/1944328/-/ji1lwz/-/index.html.
27. Hagos Gebresllassie, Temesgen Gashaw, and Abraham Mehari, "Wetland Degradation in Ethiopia: Causes, Consequences and Remedies," *Journal of Environment and Earth Science* 4, no. 11 (2014), www.iiste.org.
28. Patrick Gerland, et. al., "World Population Stabilization Unlikely This Century," *Science*, October 10, 2014, 234–237.
29. Lisa Palmer, "Famine Is a Feminine Issue," *Slate*, April 10, 2014, http://www.slate.com/articles/health_and_science/feed_the

_world/2014/04/educate_women_and_save_babies_how_to _control_population_and_end_hunger.html.

30. Uganda Bureau of Statistics, *The 2002 Uganda Population and Housing Census: Population Dynamics* (Kampala: Uganda Bureau of Statistics, October 2006), http://www.ubos.org/onlinefiles /uploads/ubos/pdf%20documents/2002%20CensusPopndynamicsAnalyticalReport.pdf.
31. Robert Engelman, "Long in Background, Population Becoming a Bigger Issue at Climate Change Discussions," *New Security Beat* (blog of the Wilson Center), November 10, 2015, https://www .newsecuritybeat.org/2015/11/population-bigger-issue-climate -change-discussions/.
32. Gordon Conway, *One Billion Hungry: Can We Feed the World?* (Ithaca, NY: Cornell University Press, 2012).
33. "2015 Global Hunger Index," International Food Policy Research Institute, October 12, 2015, http://ghi.ifpri.org/results/.
34. "What the World Eats," *National Geographic*, http://www .nationalgeographic.com/what-the-world-eats/ (accessed August 19, 2016).
35. Searchinger, ibid.

3: Soils, Sylvan Pastures, and Sustainability

1. C. Giraldo, F. Escobar, J. D. Chara, and Z. Calle, "The Adoption of Silvopastoral Systems Promotes the Recovery of Ecological Processes Regulated by Dung Beetles in the Colombian Andes," *Insect Conservation and Diversity* 4 (2011): 115–122, doi: 10.1111/ j.1752-4598.2010.00112.x.
2. Giraldo et al., "Adoption of Silvopastoral Systems."
3. E. Murgueitio et al., "Native Trees and Shrubs for the Productive Rehabilitation of Tropical Cattle Ranching Lands," *Forest Ecology and Management* 261 (2011): 1654–1663.
4. Food and Agriculture Organization of the United Nations, *Livestock's Long Shadow: Environmental Issues and Options* (Rome: FAO, 2006), http://go.nature.com/BFrtHv.
5. R. R. Vera, *Country Pasture/Forage Resource Profiles: Colombia* (Rome: Food and Agriculture Organization of the United Nations, 2006), http://go.nature.com/54etdI.
6. Julian Chará, interview by author, February 2014, Cali, Colombia.
7. Department of Energy & Climate Change, "UK Announces New Climate Programmes in Africa, South America and Other Vul-

nerable Countries through the International Climate Fund," British government press release, December 4, 2012, http://go.nature.com/fZQwMd.

8. J. A. Foley et al., "Solutions for a Cultivated Planet," *Nature* 478 (October 20, 2011): 337–342, http://www.nature.com/nature/journal/v478/n7369/full/nature10452.html.
9. T. Searchinger, *Creating a Sustainable Food Future: Interim Findings* (Rome: Food and Agriculture Organization of the United Nations, 2013).
10. Timothy Searchinger, interviews by author, June 2013 and February 2014, South Royalton, Vermont, and Princeton, New Jersey.
11. Fabiola Vega, interview by author, February 2014, Colombia.
12. Irene Montes Londoño, "Pinzacuá Farm: Case Study of Sustainable Cattle Farming in the Colombian Coffee Region," United Nations Development Program, 2012, http://www.undp.org/content/dam/undp/library/Environment%20and%20Energy/Green%20Commodities%20Facility/UNDP%20Pinzacuá%20case%20study%20(1).pdf.
13. World Bank, "Mainstreaming Sustainable Cattle Ranching," Implementation, Status and Results Report, Public Disclosure Copy, July 13, 2015, http://www-wds.worldbank.org/external/default/WDSContentServer/WDSP/LCR/2015/07/13/090224b082fef37f/1_0/Rendered/PDF/Colombia000Mai0Report000Sequence011.pdf.
14. Estella Dominguez, interview by author, February 2014, Colombia.
15. Olimpo Montes, interview by author, February 2014, Colombia.
16. Enrique Murgueitio, interview by author, February 2014, Colombia.
17. Searchinger interviews.
18. Country Profile: Colombia, USAID Land Tenure and Property Rights Portal, www.usaidlandtenure.net/colombia. See also USAID Country File, "Property Rights and Resource Governance: Colombia," http://www.land-links.org/wp-content/uploads/2016/09/USAID_Land_Tenure_Colombia_Profile.pdf.
19. Amy Lerner, phone interview with author, Princeton, New Jersey, February 2014.
20. P. Havlik et al. "Climate Change Mitigation through Livestock System Transitions," *Proceedings of the National Academy of Sciences* 111 (2014): 3709–3714; M. Herrero et al., "Biomass Use, Production, Feed Efficiencies, and Greenhouse Gas Emissions

from Global Livestock Systems," *Proceedings of the National Academy of Sciences* 110 (2013): 20888–20893.

21. Paul West, interview by author, Baltimore, Maryland, August 13, 2015; Paul C. West et. al., "Leverage Points for Improving Global Food Security and the Environment," *Science*, July 18, 2014, http://www.sciencemag.org/content/345/6194/325.full.
22. Murguietio interview. See also Rodolfo Dirzo, H. S. Young, H. A. Mooney, and G. Ceballos, *Seasonally Dry Tropical Forests: Ecology and Conservation* (Washington, DC: Island Press, 2011).

4: California and Syria

1. Abrahm Lustgarten, "Killing the Colorado," *ProPublica*, June 16, 2015, https:// www.propublica.org/series/killing-the-colorado.
2. Noah S. Diffenbaugh, Daniel L. Swain, and Danielle Touma, "Anthropogenic Warming Has Increased Drought Risk in California," *Proceedings of the National Academy of Science* 112 (2015): 3931–3936, http://www.pnas.org/content/112/13/3931; Kat Kerlin, "Drought Costs California Agriculture $1.84B and 10,100 Jobs in 2015," UCDavis.com, August 18, 2015, https://www.ucdavis.edu/news/drought-costs-california-agriculture-184b-and-10100-jobs-2015.
3. "NASA: California Drought Causing Valley Land to Sink," NASA, August 19, 2015, http://www.jpl.nasa.gov/news/news.php?feature=4693.
4. National Centers for Environmental Information, NOAA, "Drought—January 2016," https://www.ncdc.noaa.gov/sotc/drought/201601.
5. Pete Spotts, "Tale of Two Droughts: What California, Syria Can Teach about Adaptation," *The Christian Science Monitor*, March 3, 2015, http://www.csmonitor.com/Environment/2015/0303/Tale-of-two-droughts-What-California-Syria-can-teach-about-adaptation-gap.
6. John Wendle, "The Ominous Story of Syria's Climate Refugees," *Scientific American*, December 17, 2015, http://www.scientificamerican.com/article/ominous-story-of-syria-climate-refugees/.
7. Wendle, "Ominous Story of Syria's Climate Refugees."
8. Colin P. Kelley et al., "Climate Change in the Fertile Crescent and Implications of the Recent Syrian Drought," *Proceedings of the*

National Academies of Science 112, no. 11 (March 17, 2015): 3241–3246, http://www.pnas.org/content/112/11/3241.full.

9. Peter H. Gleick, "Water, Drought, Climate Change, and Conflict in Syria," *Weather, Climate, and Society* 6 (July 1, 2014): 331–340, http://journals.ametsoc.org/doi/abs/10.1175/WCAS-D-13-00059.1.
10. Kelly, "Climate Change in the Fertile Crescent."
11. F. Salamini et al., "Genetics and Geography of Wild Cereal Domestication in the Near East," *Nature Reviews Genetics* 3, no. 6 (June 2002): 429—441, http://www.nature.com/nrg/journal/v3/n6/full/nrg817.html.
12. David Rosenberg, "Food and the Arab Spring," Rubin Center, October 27, 2011, http://www.rubincenter.org/2011/10/food-and-the-arab-spring/.
13. Samir Suweis et al., "Resilience and Reactivity of Global Food Security," *Proceedings of the National Academy of Sciences* 112, no. 22 (June 2, 2015): 6902–907, www.pnas.org/cgi/doi/10.1073/pnas.1507366112; D. K. Ray et al., "Recent Patterns of Crop Yield Growth and Stagnation," *Nature Communications* 3 (2012): 1293.
14. Suweis et al., "Resilience and Reactivity of Global Food Security."
15. Ibid.
16. Global Information and Early Warning System, "Egypt: Food Security Snapshot," Food and Agriculture Organization of the United Nations, August 2, 2016, http://www.fao.org/giews/countrybrief/country.jsp?code=EGY.
17. Brian Gardner, *Global Food Futures: Feeding the World in 2050* (New York: Bloomsbury Academic, September 2013).
18. Stephen D. Simpson, "Top Agricultural Producing Countries," *Investopedia*, http://www.investopedia.com/financial-edge/0712/top-agricultural-producing-countries.aspx.
19. Emmy Simmons, interviews by author, summer and fall 2014, Washington, D.C.
20. Emmy Simmons, "Harvesting Peace: Food Security, Conflict, and Cooperation," Environmental Change & Security Program report 14, no. 3 (2013), Woodrow Wilson International Center for Scholars, Washington, DC, https://www.wilsoncenter.org/sites/default/files/HarvestingPeace.pdf,
21. Julian Cribb, *The Coming Famine* (Berkeley: University of California Press, 2010), 4.

22. Dan Charles, "American Farmers Say They Feed the World, But Do They?" *National Public Radio*, September 17, 2013, http://www.npr.org/sections/thesalt/2013/09/17/221376803/american-farmers-say-they-feed-the-world-but-do-they.
23. Iowa Corn, "Ethanol," http://www.iowacorn.org/en/ethanol/; Economic Research Service, "Corn Trade," U.S. Department of Agriculture, http://www.ers.usda.gov/topics/crops/corn/trade.aspx. Although the United States is the world's largest corn exporter, exports account for a relatively small share of demand for U.S. corn—less than 15 percent.
24. Economic Research Service, "California Drought: Crop Sectors," U.S. Department of Agriculture, May 3, 2016, http://www.ers.usda.gov/topics/in-the-news/california-drought-farm-and-food-impacts/california-drought-crop-sectors.aspx.
25. Kirk Siegler, "California Farmers Gulp Most of State's Water, But Say They've Cut Back," *National Public Radio*, April 7, 2015, http://www.npr.org/sections/thesalt/2015/04/07/398106067/calif-s-farmers-gulp-most-of-states-water-but-say-theyve-cut-back.
26. Josué Medellin-Azuara, telephone interview by author, April 22, 2015.
27. Public Policy Institute of California, Water Policy Center, "Water for Farms," http://aic.ucdavis.edu/publications/PPIC.pdf, April 2015. See also U.S. Department of Agriculture, "2014 Crop Year Report," California Agriculture Production Statistics, https://www.nass.usda.gov/Statistics_by_State/California/Publications/California_Ag_Statistics/2013/2013cas-all.pdf.
28. John Matthews, telephone interview by author, April 23, 2015.
29. Katharine Hayhoe, email to author, April 24, 2015.
30. Ben Jervey, "Exporting the Colorado River to Asia, through Hay," *National Geographic*, January 23, 2014, http://news.nationalgeographic.com/news/2014/01/140123-colorado-river-water-alfalfa-hay-farming-export-asia/.
31. University of California, Davis, Agricultural Sustainability Institute, "What Is Sustainable Agriculture?" n.d., http://asi.ucdavis.edu/programs/sarep/about/what-is-sustainable-agriculture.
32. Suweis et al., "Resilience and Reactivity of Global Food Security."

5: Population, Health, and Environment Powerfully Working Together

1. "Live Aid Concert," History.com, http://www.history.com/this-day-in-history/live-aid-concert.

2. Alex de Waal, "Is the Era of Great Famines Over?" *The New York Times*, May 8, 2016, http://www.nytimes.com/2016/05/09/opinion/is-the-era-of-great-famines-over.html (accessed May 9, 2016).
3. de Waal, "Is the Era of Great Famines Over?"
4. "Response to El Nino Caused Drought: Appeal," ACT Alliance, http://actalliance.org/wp-content/uploads/2016/02/Response-to-El-Nino-Caused-Drought-Appeal-ETH161.pdf (accessed May 9, 2016).
5. John J. Metzler, "El Nino Hits Ethiopia, on Rebound from Socialism after Generous U.S. Aid," *worldtribune.com*, December 4, 2015, http://www.worldtribune.com/el-nino-hits-ethiopia-on-rebound-from-socialism-after-generous-u-s-aid/.
6. "Ethiopian Coffee Exports to Hit Record in 2015," Agricultural Growth Program, n.d., http://ethioagp.org/ethiopian-coffee-exports-to-hit-record-in-2015/. Ethiopia's Agricultural Growth Program is a joint effort of Ethiopia, the World Bank, and international donors, including the U.S. Agency for International Development.
7. Lester Brown, *Full Planet, Empty Plates: The New Geopolitics of Food Scarcity* (New York: W. W. Norton, 2012), 3–4.
8. Duncan Green, "25 Years after the Ethiopian Famine: What We Have Learned?" *From Poverty to Power* (blog), Oxfam, October 22, 2009, https://oxfamblogs.org/fp2p/25-years-after-the-ethiopian-famine-what-have-we-learned/.
9. *Paving the Way*, directed by Michael T. Miller, YouTube video, 11:55, posted August 7, 2015, by the Environmental Change & Security Program of the Wilson Center, Washington, DC, https://www.youtube.com/watch?v=gGw5sLp62MI.
10. Mark Tran, "Ethiopia Enlists Help of Forest Communities to Reverse Deforestation," *The Guardian*, April 15, 2013, http://www.theguardian.com/global-development/2013/apr/15/ethiopia-forest-communities-reverse-deforestation.
11. Shelley Megquier and Kate Belohlav, "Ethiopia's Key: Young People and the Demographic Dividend Policy Brief," December 2014, Population Reference Bureau, Washington, DC, http://www.prb.org/pdf15/ethiopia-demographic-dividend-policybrief.pdf.
12. Sean Peoples, "From One Generation to the Next," *New Security Beat* (blog of The Wilson Center), June 17, 2015. https://www.newsecuritybeat.org/2015/06/paving-way-ethiopias-youth-road-sustainability/.

13. Leona D'Agnes et al., "Integrated Management of Coastal Resources and Human Health Yields Added Value: A Comparative Study in Palawan (Philippines)," *Environmental Conservation* 37, no. 4 (2010): 398–409, doi: 10.1017/S0376892910000779.
14. Marti Martindale, "Empowering Women," Global Team for Local Initiatives, March 1, 2016, http://gtli.us/field-stories/.
15. Africa, "Ethiopia to Receive $50m Grant Green Climate Fund," The Citizen, November 22, 2015, http://howafrica.com/ethiopia-receives-50m-grant-from-green-climate-fund/; PHE Ethiopia Consortium, "Strategic Plan 2015–2020," 2015, http://www.phe-ethiopia.org/pdf/Stratagic_plan_all_pages.pdf;
16. Pathfinder International, sidebar to "Sustaining Health, Rights and the Environment in the Lake Victoria Basin: Technical Brief," June 2015, p. 3, http://www.pathfinder.org/publications-tools/pdfs/Sustaining_Rights_2015PHE.pdf?x=52&y=20.
17. World Food Program, "Madagascar," n.d., https://www.wfp.org/countries/madagascar/overview.
18. Caroline Savitzky, presentation at the Wilson Center, October 14, 2014, Washington, DC.
19. "What Is Population, Health and Environment?" Environmental Health, http://www.ehproject.org/phe/phe.html.
20. Robert Engelman, interview by author, October 10, 2015, Norman, Oklahoma.
21. Engelman interview.

6: India's Climate-Smart Villages

1. Mrutyunjay Swain, S. S. Kalamkar, and Kalpana M. Kapadia, *State of Gujarat Agriculture, 2011–2012* (Anand, Gujarat: Agro-Economic Research Center, 2013), 2, http://spuvvn.edu/academics/academic_centres/agro_economic_centre/research_studies/R.%20No.%20146%20State%20of%20Gujarat%20Agriculture.pdf.
2. Ravin Bhai Parmar, interview by author, September 15, 2015, Dhundi village, Gujarat, India.
3. Parmar interview.
4. Tushaar Shah, interview by author, September 15, 2015, International Water Management Institute, Anand, Gujarat.
5. Indian Agricultural Research Institute, "Accomplishments," May 19, 2015, http://www.iari.res.in/?option=com_content&view=article&id=615&Itemid=1577.

6. M. L. Jat, interview by author, September 13–14, 2015, Karnal and Taorori, Haryana, and Ludihana and Noopur Bet, Punjab, India.
7. Joginder Singh, interview by author, September 14, 2015. Noopur Bet, Punjab, India.
8. Mike Davis, *Late Victorian Holocausts: El Nino Famine and the Making of the Third World*, (New York: Verso, 2002).
9. Andrew Jarvis, telephone interview by author, September 23, 2015.
10. U.S. Agency for International Development, "History: India," September 6, 2016, https://www.usaid.gov/india/history; India Directorate of Economics and Statistics, Ministry of Agriculture, "Top Ten Wheat Producing States, 2012–2013," MapsofIndia.com, http://www.mapsofindia.com/top-ten/india-crops/wheat.html.
11. David Pimentel and Anne Wilson, "World Population, Agriculture, and Malnutrition," *Worldwatch Magazine* 19, no. 5 (September–October 2004), http://www.worldwatch.org/node/554 (accessed June 1, 2016).
12. S. Naresh Kumar et al., "Vulnerability of Wheat Production to Climate Change in India," *Inter-Research Climate Research* 59, no. 3 (2014): 173–187, http://www.int-res.com/abstracts/cr/v59/n3/p173-187/.
13. Julian Cribb, *The Coming Famine* (Berkeley: University of California Press, 2010), 3.
14. "Agriculture," National Climate Assessment, U.S. Global Change Research Program, http://nca2014.globalchange.gov; and "Global Risks 2016: Climate Change and Risks to Food Security," (also see figure 3.2.1), World Economic Forum, http://reports.weforum.org/global-risks-2016/climate-change-and-risks-to-food-security/.
15. Singh interview.
16. Singh interview.
17. Vikas Chaudary, interview by author, September 13, 2015, Tararori, Haryana, India; Chaudary, presentation to international business representatives, September 13, 2015, Tararori, Haryana, India.
18. Pramod Aggarwal, interview by author, September 17, 2015, Delhi, India.
19. "Climate Smart Villages," CGIAR, https://ccafs.cgiar.org/climate-smart-villages#.V_Z6FWPSqL4.
20. Aggarwal interview.
21. Alberto Mejia, telephone interview by author, December 16, 2014.

22. Andrew Jarvis, telephone interview by author, September 23, 2015. See also World Bank, International Center for Tropical Agriculture, "Climate Smart Agriculture in Rwanda," CSA Country Profiles Series. Washington, D.C., January 1, 2015, http://hdl.handle.net/10568/69547. And see also The CGIAR Research Program on Climate Change, Agriculture and Food Security Regions, https://ccafs.cgiar.org/regions.
23. Jarvis interview.
24. Mejia interview.
25. Oscar Perez, telephone interview by author, December 22, 2014. Cordoba, Colombia.
26. Jarvis interview.
27. Seth Shames, interview by author, November 2014, Washington, D.C.

7: Climate Stupid

1. Holger Kray, interview by author, October 10, 2015, Norman, Oklahoma; Andy Jarvis, telephone interview by author, September 23, 2015.
2. Kray interview.
3. World Wildlife Fund, "Which Everyday Products Contain Palm Oil?" http://www.worldwildlife.org/pages/which-everyday-products-contain-palm-oil
4. "Updates: 2015 Fire Season: Indonesian Fire Season Progression," November 15, 2015, http://www.globalfiredata.org/updates.html#2015_indonesia. Oliver Holmes, "Forest Fires in Indonesia Choke Much of Southeast Asia," The Guardian, October 5, 2015, https://www.theguardian.com/environment/2015/oct/05/forest-fires-in-indonesia-choke-much-of-south-east-asia.
5. Lisa Palmer, "Will Indonesia Fires Spark Reform in Rogue Forest Sector?" *Yale Environment 360,* November 5, 2016. http://e360.yale.edu/feature/will_indonesian_fires_spark_reform_of_rogue_forest_sector/2928/.
6. Forest Watch Indonesia and Global Forest Watch, *The State of the Forest: Indonesia* (Bogor: Forest Watch Indonesia, and Washington, DC: Global Forest Watch, 2002), http://www.wri.org/sites/default/files/pdf/indoforest_full.pdf.
7. Nancy Harris, Susan Minnemayer, Fred Stolle, and Octavia Aris Payne, "Indonesia's Fire Outbreaks Producing More Daily Emissions Than Entire US Economy," World Resources Insti-

tute, October 16, 2015, http://www.wri.org/blog/2015/10/indonesia's-fire-outbreaks-producing-more-daily-emissions-entire-us-economy.

8. World Bank. "The Cost of Fire: An Economic Analysis of Indonesia's 2015 Fire Crisis," Indonesia Sustainable Landscapes Knowledge; Note No. 1. Washington, D.C.: World Bank Group, http://documents.worldbank.org/curated/en/776101467990969768/The-cost-of-fire-an-economic-analysis-of-Indonesia-s-2015-fire-crisis; Harris, ibid.
9. Frances Seymour, interviews by author, October 21, 2015, Washington, D.C.; and telephone interview by author, October 28, 2015.
10. Seymour, interview by author, October 28, 2015.
11. Jim Leape, telephone interview by author, October 28, 2015.
12. Seymour interview.
13. Seymour interview, October 28, 2015.
14. Suzanne Goldenberg, "Indonesia Promises to Cut Carbon Emissions," *The Guardian*, September 24, 2015, http://www.theguardian.com/environment/2015/sep/21/indonesia-promises-to-cut-carbon-emissions-by-29-by-2030; Seymour interview, October 28, 2015.
15. Seymour interview, October 28, 2015.
16. Solange Filoso, interview by author, Annapolis, Maryland, October 27, 2015.
17. Indonesia Investments, "Palm Oil," http://www.indonesia-investments.com/business/commodities/palm-oil/item166.
18. Philip Jacopson, "Malaysian Palm Oil Gian IOI Suspended From RSPO," Mongabay.com, March 25, 2016, https://news.mongabay.com/2016/03/malaysian-palm-oil-giant-ioi-suspended-from-rspo/; Philip Jacobson, "PanEco Resigns From RSPO Over 'Sheer Level of Inaction,'" Mongabay.com, June 3, 2016, https://news.mongabay.com/2016/03/malaysian-palm-oil-giant-ioi-suspended-from-rspo/; Philip Jacobson, "RSPO Pledges Reform After NGO Exposes Shoddy Palm Oil Audits," Mongabay.com, November 20, 2015, https://news.mongabay.com/2015/11/rspo-pledges-reform-after-ngo-exposes-shoddy-palm-oil-audits/; Philip Jacobson, "Oil Palm Company Accused of Violating RSPO, IPOP Standards in Indonesia," June 11, 2015, https://news.mongabay.com/2015/06/oil-palm-company-accused-of-violating-rspo-ipop-standards-in-indonesia/; see also Palm Oil news at https://news.mongabay.com/list/palm-oil/.

19. This was the membership as of late June 2016. Its membership is listed on its website, www.rspo.org, under "Membership" and a list appears after clicking on "Find Membership."
20. Roundtable On Sustainable Palm Oil, "RSPO Statement On The Indonesian Forest Fires," www.rspo.org, December 10, 2015, http://www.rspo.org/news-and-events/news/rspo-statement-on-the-indonesian-forest-fires; Oliver Balch, "Indonesia Forest Fires: Everything You Need To Know," *The Guardian*, November 11, 2015, https://www.theguardian.com/sustainable-business/2015/nov/11/indonesia-forest-fires-explained-haze-palm-oil-timber-burning; Philip Jacobson, "PanEco Resigns From RSPO Over Sheer Level of Inaction," Mongabay.com, June 3, 2016, https://news.mongabay.com/2016/06/paneco-resigns-from-rspo-over-sheer-level-of-inaction/.
21. Greenpeace, "Indonesia's Forests: Under Fire," 2015, p. 12, http://www.greenpeace.org/international/Global/international/publications/forests/2015/Under-Fire-Eng.pdf; RSPO, "RSPO Response to Cutting Deforestation Out of the Palmer Oil Supply Chain' Report by Greenpeace," March 7, 2016, http://www.rspo.org/news-and-events/announcements/rspos-response-to-cutting-deforestation-out-of-the-palm-oil-supply-chain-report-by-greenpeace.
22. "Twenty-three Companies Fined for Causing Forest Fires Leading to Indonesia Haze," *Deutsche Welle*, December 22, 2015, http://www.dw.com/en/twenty-three-companies-fined-for-causing-forest-fires-leading-to-indonesia-haze/a-18934750.
23. Philip Johnson, "Malaysian Palm Oil Giant IOI Suspended from RSPO," *Mongabay*, March 25, 2016, https://news.mongabay.com/2016/03/malaysian-palm-oil-giant-ioi-suspended-from-rspo/.
24. Britton Stephens et al., "Weak Northern and Strong Tropical Land Carbon Uptake from Vertical Profiles of Atmospheric CO_2," *Science* 316, no. 5832 (June 22, 2007): 1732–1735, http://science.sciencemag.org/content/316/5832/1732.
25. Jeffrey Hayward, interview by author, October 10, 2015, Norman, Oklahoma.
26. Hayward interview.
27. Hayward interview.
28. Hayward interview.
29. Kray interview.

30. Kray interview.
31. Adam Withnall, "Britain Has Only 100 Harvests Left In Its Farm Soil As Scientists Warn of Growing 'Agricultural Crisis,'" *The Independent*, October 20, 2014, http://www.independent.co.uk/news/uk/home-news/britain-facing-agricultural-crisis-as-scientists-warn-there-are-only-100-harvests-left-in-our-farm-9806353.html.
32. Kray interview.

8: Rice

1. Lewis Ziska, interview by author, July 12, 2013, Beltsville, Maryland.
2. Matthew Reynolds, telephone interview by author, May 29, 2014.
3. Krishna Ramanujan, "Mine Seed Banks to Feed Tomorrow's World," *Cornell Chronicle*, July 5, 2013, http://news.cornell.edu/stories/2013/07/mine-seed-banks-feed-tomorrow-s-world.
4. Ramanujan, "Mine Seed Banks"; Steven D. Tanksley and Susan R. McCouch, "See Banks and Molecular Maps: Unlocking Genetic Potential From The Wild," *Science* 277 (August 22, 1997): 1063–1066; Susan R. McCouch, et al., "Genomics of Seedbanks: A Case Study in Rice," *American Journal of Botany* 99 (February 2012): 407–423; Susan McCouch, telephone interview by author, March 11, 2015.
5. McCouch interview.
6. McCouch interview; Julia Bailey-Serres, "Submergence Tolerant Rice: SUB1's Journey from Landrace to Modern Cultivar," *Rice* (August 4, 2010) 138–147; Kenong Xu et al., "*Sub1A* Is an Ethylene-Response-Factor-Like Gene That Confers Submergence Tolerance to Rice," *Nature* 442 (August 10, 2006): 705–708, http://www.nature.com/nature/journal/v442/n7103/full/nature04920.html.
7. McCouch interview.
8. International Rice Research Institute, "Scuba Rice: Breeding Flood Tolerance into Asia's Local Mega Rice Varieties," UKAID, Department for International Development, 2010, https://assets.publishing.service.gov.uk/media/57a08b0d40f0b652dd000a80/DFID_impact_case_study_SUB1rice_FINAL_1_.pdf.
9. McCouch interview.
10. McCouch interview. See also "Pocket K, No. 19: Molecular Breeding and Marker Assisted Selection," International Service for

the Acquisition of Agri-Biotech Applications, http://isaaa.org/resources/publications/pocketk/19/default.asp.

11. McCouch interview.
12. Mark Bittman, "GMO Labeling Law Could Stir a Revolution," *New York Times*, September 2, 2016, http://www.nytimes.com/2016/09/02/opinion/gmo-labeling-law-could-stir-a-revolution.html.

9: Hot Fish

1. Pathfinder International, *Sustaining Health, Rights, and the Environment in the Lake Victoria Basin* (Watertown, MA: Pathfinder International, June 2015), http://www.pathfinder.org/wp-content/uploads/2016/09/Sustaining_Rights_2015PHE.pdf; "Helping Women to End Sex-for-Fish Culture," *IRIN/PlusNews*, December 10, 2011, http://www.irinnews.org. *IRIN*, formerly the Integrated Regional Information Networks, is an independent nonprofit news organization.
2. Food and Agriculture Organization of the United Nations, "The State of World Fisheries and Aquaculture 2016," ReliefWeb, July 7, 2016, http://reliefweb.int/report/world/state-world-fisheries-and-aquaculture-2016.
3. Norman Myers et al., "Biodiversity Hotspots for Conservation Priorities," *Nature* 403 (February 24, 2000): 853–858, http://www.nature.com/nature/journal/v403/n6772/full/403853a0.html; Pathfinder International, "Sustaining Health, Rights, and The Environment"; M. K. Kabahenda and S.M.C. Hüsken, "A Review of Low-Value Fish Products Marketed in the Lake Victoria Region," Regional Programme Fisheries and HIV/AIDS in Africa: Investing in Sustainable Solutions, Project Report 1974, World Fish Center (Uganda), http://www1.worldfishcenter.org/wfcms/file/SF0959SID/Programme%20Coordinator/Project%20Report%201974%20-%208Dec09.pdf.
4. Indian Ocean Commission, "Nile Perch Fishery Management Plan for Lake Victoria, 2015–2019," April 2015, http://commissionoceanindien.org/fileadmin/projets/smartfish/Rapport/Report-Nile_Perch.pdf; United Nations Environment Program, "Lake Victoria: Falling Water Levels Concern for Growing Population," April 2006, http://na.unep.net/atlas/africaLakes/downloads/posters/Lake_Victoria.pdf.
5. C. Camlin, Z. Kwena, and S. Dworkin, "*Jaboya* vs. *Jakambi*: Status, Negotiation, and HIV Risks among Female Migrants in the

'Sex for Fish' Economy in Nyanza Province, Kenya," *AIDS Education and Prevention* 25, no. 3 (2013): 216–231, http://www.ncbi.nlm.nih.gov/pmc/articles/PMC3717412/; "Helping Women to End Sex-for-Fish Culture."

6. Alex Opio, Michael Muyonga, and Noordin Mulumba, "HIV Infection in Fishing Communities of Lake Victoria Basin of Uganda—A Cross-Sectional Sero-Behavioral Survey," *PLoS One* 8, no. 8 (2013), http://www.ncbi.nlm.nih.gov/pmc/articles/PMC3733779/; "Catching More Than Fish: Ugandan Town Crippled by AIDS," *New Vision*, April 23, 2014, http://www.newvision.co.ug/new_vision/news/1339940/catching-fish-ugandan-town-crippled-aids.
7. "Helping Women to End Sex-for-Fish Culture."
8. "Helping Women to End Sex-for-Fish Culture."
9. "*Jaboya* vs. *Jakambi*: Status, Negotiation, and HIV Risks among Female Migrants in the 'Sex for Fish' Economy in Nyanza Province, Kenya."
10. Clive Matunga, telephone interview by author, May 23, 2016; Pathfinder International, "Sustaining Health, Rights, and the Environment in The Lake Victoria Basin."
11. "Helping Women to End Sex-for-Fish Culture."
12. Matunga interview.
13. United Nations Environment Program, *Africa: Atlas of Our Changing Environment* (Nairobi: UNEP, 2008), http://www.unep.org/dewa/africa/africaAtlas/PDF/en/Africa_Atlas_Full_en.pdf.
14. Julian Cribb, *The Coming Famine* (Berkeley: University of California Press, 2010), 30–31.
15. "Lakes around the World Rapidly Warming," National Science Foundation press release 15-148, December 16, 2015, https://www.nsf.gov/news/news_summ.jsp?cntn_id=137121. The study the NSF press release discusses is Catherine M. O'Reilly et al., "Rapid and Highly Variable Warming of Lake Surface Waters around the Globe," *Geophysical Research Letters* 42, no. 24 (December 28, 2015): 10773–10781, http://onlinelibrary.wiley.com/doi/10.1002/2015GL066235/full.
16. Alex Opio, Michael Muyonga, and Noordin Mulumba, "HIV Infection in Fishing Communities of Lake Victoria Basin of Uganda—A Cross-Sectional Sero-Behavioral Survey," *PLoS One* 8, no. 8 (August 5, 2013), http://www.ncbi.nlm.nih.gov/pmc/articles/PMC3733779/.

17. Michael Cooperman, telephone interview by author, July 2015.
18. Cooperman interview.
19. Cooperman interview; Christopher D. Golden et al., "Nutrition: Fall in Fish Catch Threatens Human Health," *Nature* 534 (June 15, 2016): 317–320, http://www.nature.com/news/nutrition-fall-in-fish-catch-threatens-human-health-1.20074#/water.
20. Golden et al., "Nutrition"; Christopher D. Golden, interview by author, Annapolis, MD, March 10, 2016.
21. Golden interview.
22. University of Maryland, "Falling Fish Catches Could Mean Malnutrition in Developing World," June 17, 2016, http://www.umdrightnow.umd.edu/news/falling-fish-catches-could-mean-malnutrition-developing-world.
23. Rosamond L. Naylor et al., "Effect of Aquaculture on World Fish Supplies," *Nature* 405 (June 29, 2000): 1017–1024, http://www.nature.com/nature/journal/v405/n6790/full/4051017a0.html.
24. Before the partnership, which was formalized in 2009, DuPont had been involved in supplying the aquaculture industry but had no experience in aquaculture production.
25. Naylor, "Effect of Aquaculture on World Fish Supplies."
26. Golden interview.
27. T. Searchinger, *Creating a Sustainable Food Future: Interim Findings* (Rome: Food and Agriculture Organization of the United Nations, 2013), p. 99, https://www.wri.org/sites/default/files/wri13_report_4c_wrr_online.pdf.
28. Scott Nichols, interview by author, May 15, 2014, Monterey, California.
29. Nichols interview.
30. Food and Agriculture Organization, "The State of World Fisheries and Aquaculture," May 19, 2014, http://www.fao.org/documents/card/en/c/097d8007-49a4-4d65-88cd-fcaf6a969776/.
31. Nichols interview.
32. Scott Nichols, telephone interview by author, July 26, 2016.

10: Small and Sustainable

1. Scott Sibbel, interview by author, August 12, 2016, Carroll, Iowa.
2. William J. Brown, "Niman Ranch—A Natural Meat Processor Case Study," International Food and Agribusiness Management Review 3 (2000) 403–421, http://dx.doi.org/10.1016/S1096-7508(01)00057-X.

3. Stephanie Strom, “Demand Grows for Hogs That Are Raised Humanely,” *New York Times*, January 21, 2014, http://www.nytimes.com/2014/01/21/business/demand-grows-for-hogs-that-are-raised-humanely.html.
4. James M. MacDonald, Penni Korb, and Robert A. Hoppe, *Farm Size and the Organization of U.S. Crop Farming*, Economic Research Report no. 152, August 2013, U.S. Department of Agriculture, http://www.ers.usda.gov/media/1156726/err152.pdf.
5. National Agricultural Statistics Service of the U.S. Department of Agriculture and the Iowa Farm Bureau, “Number of Iowa Farms and Average Farm Size 1950–2013,” *2014 Iowa Agricultural Statistics*, September 2014, p. 10, https://www.nass.usda.gov/Statistics_by_State/Iowa/Publications/Annual_Statistical_Bulletin/2014/10_14.pdf; U.S. Department of Agriculture, “County Profile: Carroll County, Iowa,” *2012 Census of Agriculture*, https://www.agcensus.usda.gov/Publications/2012/Online_Resources/County_Profiles/Iowa/cp19027.pdf; MacDonald, Korb, and Hoppe, *Farm Size and the Organization*. McDonald, Korb, and Hoppe note on p. iii of their summary: “The report introduces a measure of midpoint acreage in which half of all *cropland acres* are on farms with more cropland than the midpoint, and half are on farms with less. Midpoint acreage is revealed to be a more informative measure of cropland consolidation than either a simple median (in which half of all *farms* are either larger or smaller) or the simple mean (which is average cropland per farm).”
6. Iowa Corn, “Corn Uses: Ethanol,” www.iowacorn.org; https://www.iowacorn.org/corn-uses/ethanol/; see also Robert Wisner, “Corn Balance Sheet,” https://www.extension.iastate.edu/agdm/crops/outlook/cornbalancesheet.pdf.
7. T. Searchinger, *Creating a Sustainable Food Future: Interim Findings* (Rome: Food and Agriculture Organization of the United Nations, 2013), https://www.wri.org/sites/default/files/wri13_report_4c_wrr_online.pdf.

Index